Arithmetic –

Integers, Fractions, Decimals

<u>Electrical and Electronic Engineering Design Series</u>

Electric Circuits – Analysis and Design

Electronic Circuit Design – with Bipolar and MOS Transistors

CMOS Circuit Design – Analog, Digital, IC Layout

Digital Design – Logic, Memory, Computers

<u>Mathematics</u>

Arithmetic – Integers, Fractions, Decimals

Books written by Nicholas L. Pappas, Ph.D.

Arithmetic
Integers, Fractions, Decimals

Nicholas L. Pappas. Ph.D.

Why I wrote this text My interest was aroused when after a serious search I could not find a comprehensive text presenting the standard algorithms. Nor did I find a text that successfully explained the *ideas* of Arithmetic. There was too much emphasis on doing trivial problems. So here we are.

A Message from the Author: I have worked continuously in the electronics industry since 1950 except for 11 semesters teaching at San Jose State University (Professor and Chair Computer Engineering 1988-1993). There I discovered my talent for teaching such as it may be. After War2 I attended Lehigh University, and then transferred to Stanford where I earned the MS degree and, while working at HP in the early 1950's, the Ph.D. EE degree. (Somehow I did not get the word and formally apply for the BS degree.) Hardware design has been my principal activity. I learned enough about assembly language, Forth, C and C++ to design the software I needed for my projects. My current activity is designing integrated circuits.

Preface

This is about the fundamental ideas of Arithmetic, the theory of Arithmetic and understanding why and how Arithmetic works. This is about effective use of the practical procedures for addition, multiplication, subtraction, and division. Practical procedures you use when doing Arithmetic.

This is about what are now standard algorithms for *integer* addition, multiplication, subtraction, and division that are recognized by the world wide mathematical community. Knowing and understanding the algorithms means one has moved beyond rote knowledge of arithmetic.

Integer division creates *fractions,* which can be converted to *decimal fractions.* We present the theory of fractions and decimals as a straightforward extension of integer arithmetic.

The basic laws defining operations are presented in the last chapter in order to avoid piling on new information in earlier chapters. The laws make very clear the operations on numbers that are permissible, and why. Studying the laws reviews the entire subject.

In this text know that *elementary algebra* is used for general explanations such as *if n is a number then n+1 is the next number,* and specific numbers are used in examples. And, instead of taking up many pages with arithmetic problems, the reader is asked to select pairs of numbers to add, multiply, subtract and divide. The results can be verified by using a calculator. However fraction and decimal problems, and their solutions, are included.

The Standard Arithmetic Algorithms The word standard implies that we can order a document from a recognized Arithmetic Standards organization. We cannot do that, because we have not found such an organization. Nevertheless the world wide mathematical community recognizes what have evolved into standard algorithms for *integer* addition, multiplication, subtraction, and division. There are minor variations from country to country that are of no significance, because the underlying mathematical ideas are the same.

Arithmetic

We describe and fully explain the standard algorithms for addition, multiplication, subtraction, and division. The explanations emphasize ideas and procedures that *always* produce a solution. Perhaps you will agree with us when we say these algorithms are extraordinary discoveries.

> *An algorithm is a procedure, requiring no creative skills of the user, with precise instructions, specifying a finite number of steps, so that sooner or later the procedure ends.*

A specific virtue of the arithmetic algorithms is that they solve an N digit problem one digit at a time. Repeat: one digit at a time. In other words one N-digit problem becomes N one-digit problems (one 5-digit problem becomes five 1-digit problems).

This is important, because one-digit problems are done in one's mind.

Algorithms are used, because they are methods that show how to solve every possible problem. Algorithms *always* produce a solution. We believe knowing how to apply the algorithms means one *understands* what arithmetic is about. Using the algorithms with understanding enhances your mathematical skills. Progress is subtle, and real. Know this about the relationship of practical procedures to the algorithms.

> *The practical procedures implementing the standard algorithms use the algorithm's steps in a subtle way in order to be efficient. Consequently the procedures seem to be very different from the algorithms. They are not.*

Who can benefit from reading this text? Anybody who wants to be effective when doing Arithmetic. You may be a student who suspects he/she is being short changed by the system. You may be a school teacher, not trained in math, who is assigned to teach Arithmetic. You may be a parent, concerned about what is not taught in school, who is willing to make the effort to introduce these ideas to your children. You may be a person who wants to improve your math capability. Perhaps *who* is anyone who wants to *know*, and who wants to *be able to do*.

The text In this modest mathematics text, we have tried hard to write in plain English. We do not use the phrase *this is obvious* for a good reason. Nothing is obvious to a person learning any subject.

Why is Mathematics perceived as difficult? Well it is. Mathematics is difficult perhaps, but not impossible as some would have you believe. *Math is difficult* is an often repeated statement whose implications are simply misleading. A statement that is a disservice to anyone who is turned away from math by it, because knowing math gives you a highly competitive advantage, a real edge.

Please consider why *learning mathematics is not different nor more difficult than learning to read.*

First of all *anything* is hard if an effort is required to learn about it. Do not pick on math. For example, some of us have forgotten the long and hard effort required to learn to read. Today we just read. Nothing to it. Right? Really?

Reading a novel or newspaper is not difficult, because there is nothing else to learn except for a new word or two. You just read. The only symbols used are the letters of the alphabet and, perhaps one or two numbers. This is why reading a page may take only a few minutes. What you see on each page is familiar. However, if you do not know the meaning of specific words you may stop to look them up in a dictionary, or you simply gloss over them with essentially *no negative consequences.*

On the other hand reading mathematics is significantly different. In addition to the alphabet, there are lots of numbers, many new words, strange symbols, drawings, and perplexing combinations of letters, numbers and other symbols. The mathematics reader has to learn many things at the same time such as vocabulary, symbols, and how to manipulate mathematical expressions. Furthermore, glossing over has serious negative consequences. This complexity creates the perception mathematics is difficult. As we have said, it is. Nevertheless, let us be fair.

Accurate recall will remind you that learning to read did not happen overnight (how many years was it?). In the beginning you had to learn the alphabet symbols, the sound for each symbol, how to pronounce combinations of symbols (words), and on and on. It was not easy. However you learned by doing. The more you read the better you could read.

> *We repeat: This is why learning mathematics is not different nor more difficult than learning to read.*

Nothing new By the way there is *nothing new* about the teaching-of-math problem. Frank H. Hall wrote the monograph "Arithmetic: How to Teach it." This was published in 1900, repeat 1900. Mr. Hall's excerpts from the Report of the Committee of Ten:

> The "Committee of Ten" was appointed at the meeting of the National Education Association in Saratoga in July 1892. Its chairman was Charles W. Eliot, President of Harvard University.
>
> The conference [on Mathematics] consisted of one government official, a university professor, five professors of mathematics in as many colleges, two teachers of mathematics in endowed schools, and one proprietor of a private school for boys. The professional experience of these gentlemen and their several fields of work were various, and they came from widely separated parts of the country; yet they were unanimously of opinion that **a radical change in the teaching of arithmetic was necessary** (writer's bold).
>
> > Reference: Hall, Frank H. Arithmetic: How to Teach it. New York, Chicago, Werner School Book Company, 1900. p. cm. LC CALL NUMBER: QA135 .H2

Our blog *npappasee.blogspot.com* may offer you additional information. Take a look.

We would appreciate receiving your comments and views on this text at npappasz@yahoo.com.

Contents

1 Why we count the way we do

We count the way we do, because numbers are based on the ideas of position and position weight.

Knowing how to count is a prerequisite to any mathematical activity. Counting is ground zero. A major goal is *to be able to* write down a number, which represents any value you choose, *and*, at the same time *to know why* numbers are written the way they are. Knowing why means you understand the ideas contributing to the rules of the number system we use every day.

Learning to count requires knowing the symbols for numbers as well as the names of numbers. At first, mathematics is difficult because there are all of those new symbols and words to learn before you can really get going.

Any number represents some quantity of units. Know that the unit can be apples, cars, or whatever. The house is 27 feet wide. No, no the house is 9 yards wide. However this is about arithmetic so we choose one, one of anything, (whose symbol is 1) as our unit. Thus any number represents some quantity of ones. The symbol 5 represents the sum of five ones. Mathematics benefits immensely from abstractions such as 1, without committing to 1 whatever (cars, feet, yards).

If you select any number, *the next number is found by adding 1.* In the abstract language of algebra, if letter n is shorthand for the word number and n represents any number, then the next number is n+1.

To enable the discussion, we need to define *equals* (=) and *expression*. The = symbol connects two expressions such as A=B. The = symbol means that the expression A on the left hand side has the same value as the expression B on the right hand side.

A mathematical expression is a group of symbols, numbers and/or operators, representing a quantity. In fact we can say that, in arithmetic, everything is an expression, and that not all expressions are equal.

Note: Sooner or later we learn to count from zero to infinity. Infinity is an alias for *a number that is as large as you please.*

1.1 Zero to Nine: 0 to 9

Over the centuries people have agreed on names for numbers. There is, however, a very big problem with names, which are just words. The very big problem is that we cannot use words *to do* arithmetic, because that would be tedious and probably impossible.

We can do arithmetic if we use symbols in combinations. A study of the history of number reveals that the world has agreed on (arbitrary) names and symbols for quantities ranging from zero to nine.

zero	*one*	*two*	*three*	*four*	*five*	*six*	*seven*	*eight*	*nine*
0	1	2	3	4	5	6	7	8	9

Number Definitions Definitions are implemented by adding 1 to produce the next number.

$0 = 0$

$1 = 1$

$2 = 1 + 1$

$3 = 1 + 1 + 1$

$4 = 1 + 1 + 1 + 1$

$5 = 1 + 1 + 1 + 1 + 1$

$6 = 1 + 1 + 1 + 1 + 1 + 1$

$7 = 1 + 1 + 1 + 1 + 1 + 1 + 1$

$8 = 1 + 1 + 1 + 1 + 1 + 1 + 1 + 1$

$9 = 1 + 1 + 1 + 1 + 1 + 1 + 1 + 1 + 1$

A *number line* is a geometrical representation of number. The very important number line is a graphic display of numbers.

```
0   1   2   3   4   5   6   7   8   9   ?
└─┴─┴─┴─┴─┴─┴─┴─┴─┴─┴─┴─┴─┴─┴─┴─┴─┴─┴─┘
```

The number line is constructed by marking off equal lengths along the line. Each mark on the number line is assigned a number. Assign 0 to any mark. Next, assign 1 to the first mark to the right of zero. Then the *distance* from 0 to 1 *represents* 1 unit of length. Subsequent marks to the right add 1 unit to the distance. Label subsequent marks 2, 3, 4, and up to 9. The question mark we put on the line is discussed in upcoming paragraphs.

1.2 Counting past 9

When we try to count units beyond 9 we run out of number symbols, because the world decided not to introduce new symbols. It did not want to repeat the disaster of the Roman number system (such as XLVII or 47).

When we count units, all we can do with units is to count from 0 to 9 over and over again. We *recycle* through 0, 1, 2, 3, ..., 9, 0, 1, 2, etc. Each time we reach 9 we have counted an additional 9+1 units.

This number 9+1 needs a name, and just how do we represent 9+1? Well, the number 9+1 has to be one *whatever*, and, if we continue counting. the next 9+1 makes it two whatevers. The whatevers have to be counted by another digit taken from 0 to 9, because no new symbols are allowed.

Another digit would allow us to add 1 to it each time the unit's count recycled from 9 back to 0 (another whatever). This other digit records the number of 9+1's. This digit needs a name. Someone must have said abracadabra and arbitrarily named the next number 9+1 *ten*, and so the other digit is the *ten's* digit. No new symbols means using only 0 to 9 as the ten's digit, All digits are 0 when we start counting.

```
0123456789  0123456789  0123456789  0123456789  012....  units
0000000000  1111111111  2222222222  3333333333  444....  tens
```

As we count, units increment from 0, and each time the units recycle from 9 back to 0 to start over, the ten's digit is incremented by 1 as shown here.

The situation, however, is unsatisfactory. Here is the unit's digit and over there is the ten's digit. What is a useful way to associate the unit and ten digits, so that we can calculate with numbers greater than 9?

After a long while, there was agreement to write numbers as if they were words. Agreement to write the digits in a sequence with the highest value digit first as in 10, which one could read as 1 ten and 0 units. In this way, ten became the *two* digit number 10 (one, zero).

| Now we know the question mark ? on the number line is a 10 |

Arithmetic

We count from 0 to 9, 0 to 9, etc. Each time we count through 0, we increment the ten's digit as shown above. We continue counting until we reach 99 units, when we are faced with the question, what to do with 99+1?

We observe that 99+1 represents 9 tens + 9 units + 1 unit. We choose to convert the 9 units + 1 unit to 1 ten so that we can say 99+1 represents 9+1 or 10 tens. Consistent thinking produces yet another digit, a third digit, that counts 9+1 tens. Since names are arbitrary we say abracadabra again and the name is *hundred's digit*.

This means when we count *tens* from 0 to 9, 0 to 9, etc., we increment the hundred's digit each time we count tens through 9 to 0. An important equivalent statement is we increment the hundred's digit each time we count tens/units up to 99, and recycle through 99 to 00. When tens and units make 99, and 1 is added, there is a roll over into hundreds.

0............9	0............9	0............9	0............9	0.......	*units*
0123456789	0123456789	0123456789	0123456789	012....	*tens*
0000000000	1111111111	2222222222	3333333333	444....	*hundreds*

The first time we count tens/units through 99 to 00, or 099 to 000, we increment the hundred's digit past 0. Consequently we write 99+1 in word format as 100 (one, zero, zero).

Can you guess what's next? We continue to count and soon we reach 999. The next number is 999+1. The name is thousand, and we write it in word format as 1000 (one, zero, zero, zero). When units, tens, and hundreds equal 9, then the next +1 causes a roll over into the next thousand.

0............9	0............9	0............9	0............9	0.......	*units*
0............9	0............9	0............9	0............9	0.......	*tens*
0123456789	0123456789	0123456789	0123456789	012....	*hundreds*
0000000000	1111111111	2222222222	3333333333	444....	*thousands*

In this way, counting units forces a counting of tens. And, in the same way, counting tens and units forces a counting of hundreds. And, again in the same way, counting hundreds and tens and units forces a counting of thousands. We are on our way to infinity.

1.3 Money

Everybody knows about money. In the United States the coins and paper money are the following.

penny	one cent (cent is an alias for penny)
nickel	five cents
dime	ten cents
quarter	two dimes and one nickel
half dollar	five dimes
one dollar bill	ten dimes
five dollar bill	five one dollar bills
ten dollar bill	ten one dollar bills
and so forth	

Let us set aside the coins and paper money that are not in units of one, ten, etc. Now we have a short list.

penny	one cent
dime	ten cents
one dollar bill	ten dimes
ten dollar bill	ten one dollar bills
and so forth	

A bank is willing to exchange
   ten pennies for one dime
   ten dimes for one dollar
   ten dollar bills for one ten dollar bill
   and so forth

Read the symbol = as *equals* or *are equal to*. So we can say
   10 pennies = 1 dime   (ten p are equal to 1 d)
   10 dimes = 1 dollar
   10 dollar bills = 1 ten dollar bill
   and so forth

1.4 Ten to Nineteen: 10 to 19 -- The Teens

The new word *ten* is another name for a quantity (of pennies for example). If you take the ten pennies to the bank you can exchange them for one dime, because one dime is equivalent to ten pennies. If you do this, then you have one dime and no pennies. In arithmetic, however, we prefer to say you have 1 dime and 0 pennies. This is consistent with the number ten written as 10 (one, zero). This is what is important: *the symbol 10 for ten written as 1 and 0 side by side means one ten plus zero ones.*

When *six* more pennies are given to you, you will have *ten* and *six* objects in the form of one dime and six pennies. Abracadabra and we produce *sixteen* as the new name for the number one six (16). The numbers ten to nineteen are defined by simply adding one unit to the current number.

name	number of units	symbol
ten	11111 11111	10
eleven	11111 11111 1	11
twelve	11111 11111 11	12
thirteen	11111 11111 111	13
fourteen	11111 11111 1111	14
fifteen	11111 11111 11111	15
sixteen	11111 11111 11111 1	16
seventeen	11111 11111 11111 11	17
eighteen	11111 11111 11111 111	18
nineteen	11111 11111 11111 1111	19

Notice that every number from ten to nineteen has ten marks plus zero to nine marks. Consistent with counting from 0 to 9, 0 to 9, etc., the numbers ten to nineteen are written in the form 10, 11, 12, 13, 14, 15, 16, 17, 18, and 19. We do not know why the names of numbers 10 to 19 are not ten, ten one, ten two, etc., instead of the special words ten, eleven, ... , nineteen.

1.5 Twenty to ninety nine: 20 to 99

You have 19 pennies. The group with size nineteen has one group of ten marks and one group of nine marks, which we group by five's for clarity.

nineteen 11111 11111 11111 1111 19

Take another penny. We use the magic word abracadabra and say you now have *twenty* pennies.

Exchange the first ten pennies for one dime. Now you have one dime and ten pennies. Exchange the second ten pennies for another dime. Now you have two dimes and zero pennies.

If the value of one dime is represented by the number 10, then the value of two dimes is represented by the number 20. Think 1, 2 in the ten's position. When you count by tens the sequence of numbers is 00, 10, 20, 30, 40, 50, 60, 70, 80, 90. In other words we count from 0 to 9 in the ten's position. No new symbols are used.

You count up from twenty in a regular manner (no more funny teen names): twenty, twenty one, twenty two, and so forth up to twenty nine. Here is 23 and the numbers 20 to 29.

twenty - three 11111 11111 11111 11111 111 23
    20 21 22 23 24 25 26 27 28 29

To emphasize we rewrite 23 and 29 by replacing marks with 10.

twenty-three   10 10 111   23
twenty-nine   10 10 11111 1111   29

Take another penny. Now you have two dimes and ten pennies. This quantity is equivalent to *thirty* pennies. (We will not say abracadabra any more when new words are introduced, because by now you know they are arbitrarily selected.)

Arithmetic

As pennies continued to be passed to you, the quantity of pennies increased from twenty to thirty so that your two dimes and zero pennies increased to two dimes and ten pennies. Exchange the ten pennies for another dime. Now you have three dimes and zero pennies.

Thirty represents three tens, so we write it as 30. Remember 20? The digit 3 has *weight* ten. This is why 30 represents 3 tens and 0 ones. Here are the numbers thirty one to thirty nine. Forty to forty nine follow the thirties.

30 31 32 33 34 35 36 37 38 39

40 41 42 43 44 45 46 47 48 49

When we count by tens you might think we would say ten, one ten, two tens, and so forth up to nine tens. Well, life is not straightforward, because the straightforward was modified to count by tens as follows, which is also shown on a number line.

zero	0	*fifty*	50
ten	10	*sixty*	60
twenty	20	*seventy*	70
thirty	30	*eighty*	80
forty	40	*ninety*	90

```
0   10  20  30   40  50  60  70  80   90   ??
L___I___I___I____I___I___I___I___I____I____I____I____I____I____I____I____I
```

You simply continue this process to reach 99, which is nine tens and nine ones (ninety nine). Now you have nine dimes and nine pennies.

> Clearly the mark ?? on the number line is a 100

Zero to ninety nine: 0 to 99 – A vocabulary

0 zero	5 five
1 one	6 six
2 two	7 seven
3 three	8 eight
4 four	9 nine

10 ten	20 twenty	30 thirty
11 eleven	21 twenty one	31 thirty one
12 twelve	22 twenty two	32 thirty two
13 thirteen	23 twenty three	33 thirty three
14 fourteen	24 twenty four	34 thirty four
15 fifteen	25 twenty five	35 thirty five
16 sixteen	26 twenty six	36 thirty six
17 seventeen	27 twenty seven	37 thirty seven
18 eighteen	28 twenty eight	38 thirty eight
19 nineteen	29 twenty nine	39 thirty nine

40 forty	50 fifty	60 sixty
41 forty one	51 fifty one	61 sixty one
42 forty two	52 fifty two	62 sixty two
43 forty three	53 fifty three	63 sixty three
44 forty four	54 fifty four	64 sixty four
45 forty five	55 fifty five	65 sixty five
46 forty six	56 fifty six	66 sixty six
47 forty seven	57 fifty seven	67 sixty seven
48 forty eight	58 fifty eight	68 sixty eight
49 forty nine	59 fifty nine	69 sixty nine

70 seventy	80 eighty	90 ninety
71 seventy one	81 eighty one	91 ninety one
72 seventy two	82 eighty two	91 ninety two
73 seventy three	83 eighty three	93 ninety three
74 seventy four	84 eighty four	94 ninety four
75 seventy five	85 eighty five	95 ninety five
76 seventy six	86 eighty six	96 ninety six
77 seventy seven	87 eighty seven	97 ninety seven
78 seventy eight	88 eighty eight	98 ninety eight
79 seventy nine	89 eighty nine	99 ninety nine

1.6 One Hundred and up

The number following 9 is 9+1, which was given the name ten and written as the word 10. The number following 99 is 99+1, which is ten tens.

At this point you have nine dimes and nine pennies. You receive one more penny. Now you have nine dimes and ten pennies. You can exchange the ten pennies for another dime so that you have ten dimes and zero pennies. You can make one more exchange: ten dimes for one dollar. Right? Now you have one dollar, zero dimes, and zero pennies.

We write nine zero, 9 0, as 90 for nine tens, so write ten zero, 10 0, as 100 for ten tens. No reason not to do so. Now we have a three digit number. The name for one zero zero, 100, is one hundred: a hundred has two zeros (zero dimes, zero pennies) whereas a ten has one zero (zero pennies).

There are other ways to arrive at 100. You count up to 099 and add one more. What happens? You reset 99 to 00 and advance the hundred's digit by one. The result is 100. In other words, 099 reset to 000 and delivered a 1 from the ten's digit to the hundred's digit to get 100. Delivering that 1 is called a carry. Carry that one from this position to the next position at the left. Again in more detail.

You count up to 99 and add one more. This resets the ones digit 9 to 0 and a carry of 1 advances the ten's digit 9 by one. This forces a reset of the ten's digit 9 to 0 and a carry of 1 advances the hundred's digit 0 by one. The result is 100. A graphic description is as follows. This process is explained in the chapter on Addition.

```
11    ← carries
099
+1
———
100
```

Counting up from 100 proceeds by adding one unit at a time. Counting from 100 to 199 is the same as counting from 00 to 99 with a 1 as the third digit. The 1 is the most significant digit, because its value, one hundred, is greater than the value of any digit in the positions to the right. In 199 the value of the 1 in the hundred's position is 100, whereas the value of the 9 in the ten's position is 90, and the value of the 9 in the ones position is 9. This is why the value of the 1 in 100 is greater than the sum of the values of the digits in positions to the right of the 1. This is always true.

Next, observe that merging one hundred (100) with sixty eight (68) makes one hundred sixty eight (168). Now we can write the numbers from 100 to 199 and greater.

100	101	102	103	104	105	106	107	108	109
110	111	112	113	114	115	116	117	118	119
120	121	122	123	124	125	126	127	128	129
130	131	132	133	134	135	136	137	138	139
140	141	142	143	144	145	146	147	148	149
150	151	152	153	154	155	156	157	158	159
160	161	162	163	164	165	166	167	168	169
170	171	172	173	174	175	176	177	178	179
180	181	182	183	184	185	186	187	188	189
190	191	192	193	194	195	196	197	198	199
200	201	and so forth

Observe that 100 plus 99 plus 1 = 100 plus 100. Reminder: 10 plus 10 equals two tens or 20, so 100 plus 100 = two hundreds or 200.

Next we count up from 200 to 299. Adding one to 299 the next number must be three hundreds, 300. And so we can count up to 999 (nine hundred ninety nine). In other words we count from zero in the hundred's digit to generate the 'hundreds' numbers: 000, 100, 200, 300, 400, 500, 600, 700, 800, 900. Now increase 999 by one unit. What do you get? Nine hundred ninety nine plus 1 is nine hundreds plus nine tens plus ten ones so you get ten hundreds.

We write nine zero zero, 9 0 0, as 900 for nine hundreds, so why not write ten zero zero, 10 0 0, as 1000 for ten hundreds? This makes sense, because it is consistent with 1 0 0 or 100, and 1 0 or 10. Now we have a four digit number. The name for one zero zero zero, 1000, is one thousand. Add one to get one thousand one, 1001. By now you realize that there is no end to this process.

One hundred to Nine hundred ninety nine: 100 to 999

100 one hundred
101 one hundred one
102 one hundred two
103 one hundred three
104 one hundred four
105 one hundred five
106 one hundred six
107 one hundred seven
108 one hundred eight
109 one hundred nine

110 one hundred ten
111 one hundred eleven
112 one hundred twelve
113 one hundred thirteen
114 one hundred fourteen
115 one hundred fifteen
116 one hundred sixteen
117 one hundred seventeen
118 one hundred eighteen
119 one hundred nineteen

120 one hundred twenty
121 one hundred twenty one
122 one hundred twenty two
123 one hundred twenty three
124 one hundred twenty four
125 one hundred twenty five
126 one hundred twenty six
127 one hundred twenty seven
128 one hundred twenty eight
129 one hundred twenty nine

130 one hundred thirty
131 one hundred thirty one
132 one hundred thirty two
133 one hundred thirty three
134 one hundred thirty four
135 one hundred thirty five
136 one hundred thirty six
137 one hundred thirty seven
138 one hundred thirty eight
139 one hundred thirty nine

140 one hundred forty
141 one hundred forty one
142 one hundred forty two
143 one hundred forty three
144 one hundred forty four
145 one hundred forty five
146 one hundred forty six
147 one hundred forty seven
148 one hundred forty eight
149 one hundred forty nine

150 one hundred fifty
151 one hundred fifty one
152 one hundred fifty two
153 one hundred fifty three
154 one hundred fifty four
155 one hundred fifty five
156 one hundred fifty six
157 one hundred fifty seven
158 one hundred fifty eight
159 one hundred fifty nine

160 one hundred sixty to
169 one hundred sixty nine

170 one hundred seventy to
179 one hundred seventy nine

180 one hundred eighty to
189 one hundred eighty nine

190 one hundred ninety to
199 one hundred ninety nine

200 two hundred to
299 two hundred ninety nine
(same as 100 to 199, replace 1 with 2. and so forth)

900 nine hundred to
999 nine hundred ninety nine

1.7 A Short Story

Here, we show that introducing counting men is analogous to introducing new digits at the left.

In primitive times, before the invention of writing and names for numbers, counting was carried out using man's fingers without resort to languages or symbols. If a chief wanted to count the large number of animals in his herd, he would assign counting tasks to his men that went like this. The first counter, who we call man-Zero, raised one finger as each animal passed by him. After the ninth animal passed he had nine fingers raised. When the next animal passed by he closed his nine fingers to show zero fingers.

A second counter, man-One, seeing that man-Zero had closed his fists raised one finger. We want to emphasize that as man-Zero's count returned to zero, man-One's count increased by one. At this count the array of raised fingers: was this, where each 1 represents one finger.

man – One	*man – Zero*
1	0

Consider the number of animals to be 10 (one zero) corresponding to the raised fingers. This is why each finger of man-One represents 10 (one more than 9) animals, whereas each finger of man-Zero represents 1 animal.

After the ninth group of 10 (one zero) animals passed, man-One has nine fingers raised. When the next 10 group passed, man-Zero closed his nine fingers to show zero fingers. This time man-One, seeing that man-Zero had closed his nine fingers, also closed his nine fingers to return his count to zero. Now man-One is out of counting capacity. In the meantime man-Zero is raising fingers as more animals pass by.

A third counter, man-Two, was enlisted to help. His task was to raise one finger each time man-One *and* man Zero closed their nine fingers. Each finger of man-Two represents 100 animals (one more than nine groups of 10 plus 9), whereas each finger of man-One represents 10 animals. After all of the animals passed, the array of raised fingers was this.

Arithmetic

man-two	*man-one*	*man-zero*
111	11111 111	1111
3	8	4

Each counter counts from 0 to 9 over and over again. Each time a count went from the maximum 9 (or 99) back to the minimum 0 the count at the left was increased by 1. We need a new vocabulary to describe this process and the number three eight four (384), which occupies three positions. By definition the numerals three, eight, and four (3, 8, 4) occupy positions 2, 1, and 0 respectively.

A digit in position 0 represents a quantity of 1 animal.
A digit in position 1 represents a quantity of 10 animals.
A digit in position 2 represents a quantity of 100 animals.

Note that position numbers 0, 1, 2 equal the numbers of zeros following the 1 in each number. This is true for *any* number of positions.

Ten and Hundred Each man doing the counting has one more than 9 fingers. Since there was no *logical* way to deduce the name of this next number you know we agreed that the name for 10 (one zero) is ten, and the name for 100 (one zero zero) is one hundred.

A digit in position 0 represents a quantity of 1's (ones).
A digit in position 1 represents a quantity of 10's (tens).
A digit in position 2 represents a quantity of 100's (hundreds).

384 Now we can state the meaning of 384, which is
 3 hundreds plus 8 tens plus 4 ones

So? Ten counters can count all of the people on Earth without uttering a word, or writing any numbers down. However, man wants to write numbers in order to calculate, keep records, and so forth.

Consequently a system for writing and calculating with numbers was created using the symbols 0 to 9, the concept of position, and the concept of position weight.

1.8 Zero to Infinity

There is always a greater number. To any number n you add 1 to make the next number n+1.

For example, increase the group of 9999 by one unit. What do you get?

Nine thousand, nine hundred, ninety, nine, plus 1 is nine thousands, plus nine hundreds, plus nine tens, plus ten ones so you get ten thousands.

We write nine zero zero zero, 9 0 0 0 as 9000 for nine thousands. And then we write ten zero zero zero, 10 0 0 0, as 10000 for ten thousands? This is consistent with 10 0 0 or 1000, 10 0 or 100, and 1 0 or 10. Now we have a five digit number.

Add one to get ten thousand one, 10001.

Add one to 100 000 000 to get 100 000 001 (To be clear we have separated the digits into groups of three.)

Add one to 345 543 672 879 to get 345 543 672 880.

And on to infinity.

The genius of the decimal number system is that we only need the symbols 0 to 9, and the concept of position, to write any number we please, such as the numbers:

1876   314159   034   78241998543   3000000000001   4294967296

Then the concept of position weight gives each number a unique value.

1.9 Counting Binary Numbers

We have been counting with ten different symbols. This system is the decimal number system with *base* ten. A number system can have *any* base you desire. As you probably know, the base 2 (binary) number system is used by computers.

The one digit numbers used in the binary number system are zero (0) and one (1). Numbers from zero to infinity only use 0 and 1 such as 111010110. Confusion is avoided by adding subscript 2 as in 111010110_2.

When we try to count beyond 1 we run out of number symbols. We avoid new symbols when we replace the next number 1+1 by a digit that represents a quantity of 1+1 units. This new digit is named *two*. This is a $base_2$ system. The meaning of 10 (one, zero) is 2 in the $base_2$ system. So we have to know the base in order to know the value of 10 (one, zero).

The position 0 digit has weight 1 and the position 1 digit has weight 2. In the decimal system the position 0, 1, 2, 3, etc. weights are 1, 10, 100, 1000, etc. (multiples of 10). We infer that the binary system position 0, 1, 2, 3, etc. weights are 1, 2, 4, 8, etc. (multiples of 2).

Two makes sense, because we run out of symbols after two counts 0 to 1, whereas the decimal system runs out of symbols after ten counts 0 to 9. Two counts implies many digits in small binary numbers.

0	1	2	3	4	5	6	7	8	9	10	11	12	13	14	15	16
0	1	10	11	100	101	110	111	1000	1001	1010	1011	1100	1101	1101	1111	10000

Binary 1101 equals decimal 8+4+0+1=13.

The three columns of numbers show how 1, 2 and 3 digit binary numbers recycle. The first column recycles 0, 1. The second column recycles 00, 01, 10, 11 or 0 to 3_{10}. The third column recycles 000 to 111 (0 to 7_{10}).

0	00	000
1	01	001
0	10	010
1	11	011

The rule is the same as the decimal rule. Increment the digit in position n+1 (the digit to the left) whenever all of the digits to the right of position n+1 rollover from 111... and recycle back to 0.

0	00	100
1	01	101
0	10	110
1	11	111

1.10 The Ideas of Position and Position Weight

The four digit number 1876 has four digits occupying four positions. Each position can only contain one digit, which is taken from the list of one digit numbers 0, 1, 2, 3, 4, 5, 6, 7, 8, 9. The four positions are filled by 1, 8, 7, and 6 respectively to create the number 1876. In other words, numbers greater than nine must have two or more digits.

We are not limited to four positions. We can have an unlimited number of positions. The rules are that any digit from 0 to 9 can be written in any position, and there can be as many positions as we desire. This is why we can write down any numbers we please such as these using only ten different symbols. We consider this to be a remarkable number system.

1876   314159   034   78241998543   3000000000001   4294967296

The positions are assigned numbers starting with 0^1 at the right hand digit. The digits of 1876 are in positions 3, 2, 1, 0.

Now what? We need a new idea. In fact it was a *great* idea that someone unknown to us revealed a long time ago. The idea is assigning different weights for different positions in a number. (We used 1, 10, 100, ... weights when we counted past 9.)

When we count past 9 we have accumulated tens, hundreds, and so forth, which we report by incrementing the digit to the left when *all* digits to the right are 9. Each digit in a number is treated in the same way. This display shows hundreds incrementing by 1 as units and tens pass through 99 to 00.

0............9	0............9	0............9	0............9	0.......	*units (ten recycles)*
0123456789	0123456789	0123456789	0123456789	012...	*tens*
0000000000	1111111111	2222222222	3333333333	444....	*hundreds*

This is why giving each digit position a number and a weight turns out to be the key steps in the process leading to larger numbers.

---

[1] The significance of defining position zero as zero instead of one becomes very clear when you study decimals. There you learn about the idea of powers of ten and that 10_{10} to the zero power, 10^0, equals 1.

Arithmetic

Base ten What are the words describing the values of these numbers?
1876   314159   034   78241998543   3000000000001   4294967296

The weights of the positions have names. These are the base ten names. Base ten, because the weight increases ten times when we move one digit left (e.g. you add a zero to 100 to produce 1000). We include spaces for clarity when we write large numbers.

Digit name	position	position weight
ones	0	1
tens	1	10
hundreds	2	100
thousands	3	1 000
ten thousands	4	10 000
millions	6	1 000 000
billions	9	1 000 000 000
trillions	12	1 000 000 000 000

The number 2705 Now we can say the number 2705 consists of two thousands, seven hundreds, zero tens, and five ones. We read the number 2705 as two thousand, seven hundred, five. Zero digits are not voiced

Position or place Many writings use place. We prefer position to place. Our reaction is that position is more definitive than place. We feel awkward saying digit 6 is in *place* 3 in the number 116210 . This why we prefer to say digit 6 is in *position* 3 in the number 116210.

Weight and value The *weight* of position 3 is 1000, and a digit 6 in position 3 contributes the *value* 6000 to the number 116210.

We distinguish position weight from value a digit contributes. This is why we say the weight of position 3 is 1000, and the value of digit 6 in position 3 is 6×1000=6000.

1.11 Review

The Basic Ideas of Counting Start counting from zero. Count up from 0 by adding one to get the next number: zero, one, two, three, four, five, six, seven, eight, and nine. When 9 is reached adding 1 converts the 9 to 0 and a 1 is added to the left of the 0. We say 09 rolled over to become 10. This happens each time a nine is reached: e.g. 9 to 10, 39 to 40, 519 to 520, 999 to 1000. This is why we only need the ten symbols 0 to 9 to write out any number.

Position and position weight A position number is assigned to each digit in a number. The position numbers match the digits in the contrived number 9876543210. A number's right hand digit is in position zero. The digit 3 of the number 5603 is in position zero.

The position weight of position 2 is 100 so that the value of the 6 in position 2 in the number 5603 is 6×100 or 600 units. The position weight of position 3 is 1000, and so the digit 5 represents 5000 units.

Note that the number of zeros in the position weight matches the position number: For example the weight of position 9 is 1 000 000 000.

The digit in any position tells you how many position weights contribute to the number: for example the 6 in 456071 contributes six position 3 weights of 1000, and the 0 contributes zero position 2 weights of 100. This is why placing zeros to the left of a number (000456071) does not change the number's value.

One digit per position One reason this number system works is that each position is occupied by only one digit.

Only 0 to 9 in any number As we have said any digit from 0 to 9 can be written in any position, and there can be as many positions as we desire. This why we can write down any numbers we please such as these using only ten different symbols. We consider this to be a remarkable number system.

1876   314159   034   78241998543   3000000000001   4294967296

2 Addition

We could say addition is a quick way to count. If we want to know the total number of eggs in two boxes of 12 eggs we could start from 12 and count to 24, or we could add 12 to 12 to produce the result 24. We prefer to add.

The + symbol is the addition operator. The + symbol is placed before the second 12 in 12+12. This is read as twelve plus twelve. The + symbol can be used to indicate the sum of more than two numbers such as the expression 23 + 108 + 5 + 746.

As we counted (Chapter 1) we recycled the units digit from 0 to 9, 0 to 9, etc. Each time we counted through 9 we incremented the tens digit. We continued counting until we counted through 99. Consistency guided us to create and increment the hundreds digit. Then we reached 999, and we continued on towards infinity. When we use the Standard Arithmetic Algorithm these remarks are written as follows.

1	11	111	*carry*
9	99	999	*n*
+1	+ 1	+ 1	*add* 1
10	100	1000	*sum n* +1

We start by organizing the addition of any two one digit numbers, such as 6 and 3. Our work produces the Addition Table.

We are able to write any number we please when we apply the ideas of position and position weight. In other words numbers can be expanded as a sum of terms as in 3517=3000+500+10+7 and 1256=1000+200+50+6. Observe that there is one term for every digit in the numbers 3517 and 1256 with weights 1000, 100, 10, and 1.

Addition of two numbers such as these is accomplished by adding individual terms of the same weight. The *one* 4 digit addition problem is converted into *four* 1 digit addition problems (3000+1000, 500+200, 10+50, 7+6). We recognize that if we ignore the zeros these are simply one digit additions (Section 2.1). These steps are the basis of the addition algorithm.

2.1 The Addition Table

All tasks are facilitated by tools. The addition table is the basic tool used when adding numbers. This is a tool we carry around in our mind.

Numbers 0 to 9 are defined as sums of ones (Section 1.1). We can construct an addition table by counting ones. A better way starts by entering the first row of numbers 0 to 9. Then add 1 to each number to build the second row. Repeat for the third row, and so forth.

The Addition table: the sum r+c

$r\downarrow\ c\rightarrow$	0	1	2	3	4	5	6	7	8	9
0	0	1	2	3	4	5	6	7	8	9
1	1	2	3	4	5	6	7	8	9	10
2	2	3	4	5	6	7	8	9	10	11
3	3	4	5	6	7	8	9	10	11	12
4	4	5	6	7	8	9	10	11	12	13
5	5	6	7	8	9	10	11	12	13	14
6	6	7	8	9	10	11	12	13	14	15
7	7	8	9	10	11	12	13	14	15	16
8	8	9	10	11	12	13	14	15	16	17
9	9	10	11	12	13	14	15	16	17	18

The letter symbols r and c represent the words row and column. The symbols r and c in the table represent any one digit numbers. The c numbers are the numbers of the table's columns. The r numbers are the numbers of the table's rows. For example, the table entry at r row 5, c column 6 is 11, because $r + c = 5 + 6 = 11$. The highest number that appears in the table is 18, because if we count the nine 1's of the r=9 and the nine 1's of the c=9 we get a count of eighteen. That is to say nine plus nine equals 18 (9+9=18).

The addition table has 100 sums. If we are to be efficient with addition then we need to know the table. By efficient we mean fast, right now, quick. Counting on our fingers is very slow, so we do not want to do that.

2.2 Position and Position weight revisited

Position weight implements the idea of having *different weights* such as 10000, 1000, 100, 10, 1 *for the positions of digits* in numbers. A necessary condition for the position weight idea to work is that any two adjacent columns have the same increase in weight. Observe that the weight of any two adjacent positions increases *times ten* when you move left one position. If you are not ready for *times ten*, then note that each move to the left adds a zero to the weight.

Position #	8	....	4	3	2	1	0
Weight	100000000	....	10000	1000	100	10	1

Observe that the *number of zeros in the weight equals the position number.* Position 0's weight has no zeros, position 1's weight has 1 zero, and so forth.

When position weight increases from 1 to 10, 10 to 100, 100 to 1000, etc. as you move one column *to the left* of any column you are using a base 10 number system.

In position 0 the one digit numbers range from 0 to 9. The next number ten equals 9+1, and is *defined* as the two digit number 10.

In position 1 the one digit numbers also range from 0 to 9. The next number ten equals 9+1, and is *defined* as the three digit number 100.

In position 2 the one digit numbers also range from 0 to 9. The next number ten equals 9+1, and is *defined* as the four digit number 1000.

And so on.

When you say two thousand ... thirty... four... as you look at the number 2034, you are using digit value. *By definition* the value of the 3 digit in position 1 is the sum of three position 1 weights of 10, which equals 30. The value of digits 2, 0, 3, 4 is 2000, 0, 30 and 4 according to position weight. This means we can expand 2034 as a sum of terms, which is 2000 + 000 + 30 + 4.

Emphasis: only one digit is allowed in each position.

2.3 The Standard Addition Algorithm

The ideas of position and position weight in effect define the number 2034. The world says the definition is $02034 = 00000 + 2000 + 000 + 30 + 4$. The definition of 56913 is the equation $56913 = 50000 + 6000 + 900 + 10 + 3$.

Common sense tells you that you can add terms with the same weight: $00000 + 50000$, $2000 + 6000$, $000 + 900$, $30 + 10$, $4 + 3$. When we drop the zeros we have five 1 digit problems: $0 + 5$, $2 + 6$, $0 + 9$, $3 + 1$, $4 + 3$. Clearly there are five calculation steps, one step for each digit. This is what the standard addition algorithm calculates.

Standard Addition Algorithm Add 56913 to 2034. Use the equations defining the numbers. Observe that all steps are identical, and each column sum is less than 10. Section 2.4 considers column sums greater than 10.

Step 1: Execute a one digit addition in position 0.

Add the units 3+4 = 7. Enter the sum 7 in answer position 0.

```
  2034
+ 56913
     7
```

Step 2: Execute a one digit addition in position 1.

Add the tens 1+3 = 4. Enter the sum 4 in answer position 1.

```
  2034
+ 56913
    47
```

Step 3: Execute a one digit addition in position 2.

Add the hundreds 9+0 = 9. Enter the sum 9 in answer position 2.

```
  2034
+ 56913
   947
```

Step 4: Execute a one digit addition in position 3

Add the thousands 6+2 = 8. Enter the sum 8 in answer position 3.

```
  2034
+ 56913
  8947
```

Step 5: Execute a one digit addition in position 4

Add the ten thousands 5+0 = 5. Enter the sum 5 in answer position 4 (missing digits are set to 0). Done.

```
  2034
+ 56913
 58947
```

Arithmetic

Verify this result by expanding 02034 and 56913 as a sum of terms, and adding term by term.

$02034 = 00000 + 2000 + 000 + 30 + 4$
$56913 = 50000 + 6000 + 900 + 10 + 3$
the sum is
$58947 = 50000 + 8000 + 900 + 40 + 7$

The standard algorithm layout for 2034+56913, and the five one digit equations used in the algorithm are

$$\begin{array}{l} 2034 \\ +56913 \\ \hline 58947 \end{array} \quad \rightarrow \quad 0+5=5 \quad 2+6=8 \quad 0+9=9 \quad 3+1=4 \quad 4+3=7$$

The digits in any single column have the same weight. E.g. there are three 10's in 2034 and one 10 in 56913. So there are four 10's in the sum 58947.

The addition algorithm starts at the right by adding the 1's digits, moves one position left and adds the 10's digits, and so forth to produce the column sums 7, 4, 9, 8, 5 as you use the Addition Table tool.

Emphasis: omitted leading digits are 0's (2034 = 02034).

Why the algorithm starts at the right is explained in the next section.

We can rewrite the five one digit equations to emphasize position and position weight.

$$\begin{array}{ccccc} 00000 & 2000 & 000 & 30 & 4 \\ +50000 & +6000 & +900 & +10 & +3 \\ \hline 50000 & 8000 & 900 & 40 & 7 \end{array}$$

One addition of 5 digit numbers is converted to five additions of 1 digit numbers. A conversion facilitated by the ideas of position and position weight. In other words ...

The addition of each column of digits is *independent* of the other columns.

2.4 The Standard Addition Algorithm and the Carry

Addition of two one digit numbers produces two digit sums (why the zeros are important will be explained in a moment).

```
  0      4      6      9
+0     +3     +7     +9
---    ---    ---    ---
 00     07     13     18
```

The two digit sums 13 and 18 present no problems when adding one digit numbers. They do not appear to violate the one digit per position rule, because there are no other digits to the left. We did not think twice when we wrote down 13 as a two digit number 1, 3 in positions 1, 0. We avoided writing the sum of 13 ones as a two digit number in position 0. We recognized that the 13 ones are equivalent to 1 ten plus 3 ones, because ten 1's are equivalent to 1 ten. In effect we exchanged ten 1's for one 10. In money terms we went to the bank where we exchanged ten pennies for one dime. Consequently we created the two digit number 13.

Carry All column sums are two digit sums ranging from 00 to 18. The column sum 13 is expanded into 10+3. The one digit per position rule requires us to place the 3 ones in position 0, and the 1 ten in position 1. The column sum 07 is expanded into 0+7. The one digit per position rule requires us to place the 7 ones in position 0, and the 0 tens in position 1.

Since there is one ten or no ten to carry we simply enter the 1 or 0 at the top of the tens column to be added later when we sum the tens column on the left. This is what the Standard Addition Algorithm does.

> The math community says we *carry* 0 or 1 into the position at the left.

The carry allows us to satisfy the condition that *only one digit is allowed in any position so that we can add each column independently.*

The existence of a carry requires always starting addition at the right hand column where we add the position 0 digits. We cannot add the tens column (the position 1 digits) correctly until we know the carry from the ones column. The same comment applies to each position.

Standard Addition Algorithm with carry Add 56913 to 60388.

Step 1: Execute a one digit addition-with-carry in position 0.

Add the units 8+3 = 11 = 1×10 + 1
Enter the unit's 1 in answer position 0, and the ten's 1 in carry position 1.

1
60388
+ 56913
1

Step 2: Execute a one digit addition-with-carry in position 1.

Add the tens 1+8+1 = 10 = 1×10 + 0
Enter the unit's 0 in answer position 1, and the ten's 1 in carry position 2.

11
60388
+ 56913
01

Step 3: Execute a one digit addition-with-carry in position 2.

Add the hundreds 1+3+9 = 13 = 1×10 + 3
Enter the unit's 3 in answer position 2, and the ten's 1 in carry position 3.

111
60388
+ 56913
301

Step 4: Execute a one digit addition-with-carry in position 3.

Add the thousands 1+0+6 = 07 = 0×10 + 7
Enter the unit's 7 in answer position 3, and the ten's 0 in carry position 4.

0111
60388
+ 56913
7301

Step 5: Execute a one digit addition-with-carry in position 4.

Add the ten thousands 0+6+5 = 11 = 1×10 + 1
Enter the unit's 1 in answer position 4, and the ten's 1 in carry position 5.

10111
60388
+ 56913
17301

Step 6: Execute a one digit addition-with-carry in position 5.

Add the hundred thousands 1+0+0 = 01 = 0×10 + 1
Enter the unit's 1 in answer position 5, and the ten's 0 in carry position 6 (not shown). Since the carry is 0 we are done.

10111
60388
+ 56913
117301

The Standard Addition Algorithm with carry
1. Enter a 0 for each missing leading digit.
2. Start the addition process at position 0.
3. Add the column of digits.
4. Enter the sum's one's digit as the column sum. Enter the sum's ten's digit as a 0 or 1 carry to the left digit's column.
5. Move one column to the left if it exists, and go to step 3. Or, exit if there is no column to the left.

The carry can be greater than one This can occur when more than two numbers are added at the same time. Carries greater than one can occur in every multiplication problem, which is equivalent to addition of many copies of the same number (Chapter 3).

```
12212  ←carries
 99999
 19945
+49309
169253
```

Example of what not to do The non-standard layout at the left creates a second addition problem. which is why it is not the standard. The standard method is at the right.

```
   27
 + 59
 ────
   16    sum of 7 and 9
 +07     sum of 2 and 5
 ────
  086    sum of 070 and 16
```

```
  01    carry
  27
+ 59
────
 086    sum
```

Arithmetic

2.5 Review

The Basic Ideas of Addition
Addition was developed to satisfy the need to count efficiently. And so the world constructed a theory of addition that is based on the ideas of position numbers, position weights, only one digit allowed in any position, one digit additions of digits with the same weight, and the carry.

Addition is done the way it is done, because only one digit numbers 0 to 9 can reside in any position. The carry allows us to satisfy this requirement. Consequently we can add digits in any one column independently of the other columns in the many digit numbers we are adding.

Position weight and Position
The digits in any number occupy positions. The positions are assigned numbers. Starting from the right end of a number the position numbers are 0, 1, 2, ... as you move to the left. The digits in the contrived number 6543210 equal their position numbers.

Only one digit occupies each position. In any position, the digit ranges in value from 0 to 9.

Starting from the right end of a number each position has the weight 1, 10, 100, 1000, ... For a five digit (five position) number the position weights and values are as follows.

position:	4	3	2	1	0
position weight:	10000	1000	100	10	1
position values for 54328:	50000	+4000	+300	+20	+8

Adding many digit numbers - Using the Carry Idea
The addition table provides the sums of one digit numbers. The carry idea extends addition to many digit numbers.

The Standard Addition Algorithm implements addition of two many digit numbers by using a sequence of one digit additions, while applying the carry idea that exchanges ten ones for one ten. The carry allows addition of each column to be independent of other columns. This why the algorithm is a sequence of identical steps. Identical steps that convert one N digit problem to N one digit problems.

3 Multiplication

Multiplication is an efficient way to add many copies of a number. Do we want to add 375 copies of 9361? No. We prefer to multiply 375 and 9361. The new items are the multiplication table of one digit numbers, and the Standard Multiplication Algorithm. The new hurdle is the multiplication table, which we can learn in a straightforward, sensible, way that is *not* a memorization method. Addition is a prerequisite.

The multiplication operator symbols are the × , dot · , asterisk *, and parens 5(4) = 5 × 4. Multiplication is implemented by forming products of one digit numbers (2 × 7), which we refer to as level 0. At the next level (level 1) the same ideas are used to multiply a many digit number by a one digit number (5 × 9361), while using the same carry concept used in addition. Moving up to level 2 the same ideas are used again to multiply a many digit number by a many digit number (375 × 9361), which also requires the use of addition to sum what are referred to as the level 1 partial products 3×9361×100, 7×9361×10, 5×9361×1.

Addition is a part of the process, however the number of additions is reduced to one.

How do we multiply any two numbers? We will show that the product of any two numbers is found in principal
  by expansion of the multiplier as a sum of terms,
  by a sequence of one digit multiplication's with carries
  followed by one addition.

These three steps represent a way to understand what is going on, however this is not how multiplication is done in practice. Multiplication is done in practice by executing an efficient procedure based on the Standard Multiplication Algorithm.

3.1 Starting up with Multiplication

When two or more identical numbers are added together, there is another way to express their sum. For example,

z + z is the same as 2 × z
z + z + z is the same as 3 × z
z + z + z + z is the same as 4 × z

and so forth, where × *is the symbol for multiplication*. In this way we introduce the idea of multiplication. The expressions read as follows.

read 2 × z as 2 times z (*times* is an alias for *multiplied by*)
read 3 × z as 3 times z
read 4 × z as 4 times z

If any other number multiplies a number expressed by a letter, then we simply put that number before the letter.

x multiplied by 20 is expressed by 20x
y multiplied by 1 is expressed by 1y or y
z multiplied by 30 is expressed by 30z

Furthermore we can multiply these products by other numbers; such as,
2 times 3x, makes 6x,
3 times 4y makes 12y,
5 times 7z makes 35z.

These products may be multiplied by yet other numbers. If a letter represents the number multiplying another number, then that letter is placed immediately before the other number. Thus, when multiplying y by x, the product is written as xy. If xy is multiplied by w, then the result is written as wxy. Changing the order in which the letters are multiplied together does not change the product. The product xy is the same as yx[1] ; y multiplied by x is the same as x multiplied by y. To understand this, we only have to substitute for x and y known numbers, e.g. 3 and 4, and the result is straightforward: 3 times 4 is the same as 4 times 3.

---

[1] Commutative law of multiplication, Chapter 8

When we substitute numbers for letters multiplied together as shown above, we must be careful not to change the meaning of the product. We cannot write them in the same way by simply placing one after the other. Consider replacing product xy with 5 and 7. If we simply write 57 for xy, then the result *as written* is the number 57, and not the product 35. When we multiply numbers such as integers, we separate them by the $\times$ sign. And so 5×7 means 5 times 7 or 35, $1 \times 2 = 2$; and $1 \times 2 \times 3 = 6$. Similarly $1 \times 2 \times 3 \times 4 \times 5 \times 6 \times 7 \times 8 \times 9 \times 10 = 3,628,800$.

We evaluate an expression such as $5 \times 7 \times 8 \times wxyz$ in the same way. First multiply 5 by 7, then multiply this product by 8 to get 280. Then we multiply 280 by w, by x, by y, and finally by z.

> The results of multiplication of two or more numbers are referred to as *products*. The individual numbers are referred to as *factors*.

Up to this point we have only worked with positive numbers. The product xy can be rewritten as $(+x) \times (+y)$, and the product is $+xy$. We can examine what results are produced by $(-x) \times (+y)$, $(+x) \times (-y)$, and $(-x) \times (-y)$. Begin by multiplying $-x$ by $+y$. Consider an analogy: if $-x$ is a debt, then y times that debt is $-xy$, or more debt. So $(-x) \times (+y)$ is the same as $-xy$, and we can say a positive quantity multiplied by a negative quantity, or vice versa, results in a negative product. Note the symbol $=$ here means "the same as."

$$(+x) \times (+y) = +xy$$
$$(-x) \times (+y) = -xy$$
$$(+x) \times (-y) = -xy$$

Now, what does $(-x) \times (-y)$ equal? The product is xy, and the sign is + or −. It cannot be the − sign; for $(-x) \times (+y)$ is the same as $-xy$, and $(+x) \times (-y)$ is the same as $-xy$. These results imply that the sign must be the opposite one to −, i.e. +. Consequently $(-x) \times (-y) = +xy$

The famous mathematician Felix Klein in his book "Elementary Mathematics from an Advanced Standpoint-Arithmetic, Algebra, Analysis" makes the point that "the rule of signs is not susceptible to proof; one can only be concerned with recognizing the logical permissibility of the rule." This is what we have just tried to do.

Arithmetic

Definitions Consider 9361×375 = 3,510,375. The number being multiplied (9361) is the *multiplicand*, 375 is the *multiplier*, 3,510,375 is the *product*, and, the numbers 9361 and 375 are *factors*.

Integers and their Factors A product is created by multiplying two or more numbers together. Then the numbers are factors of the product. For example, the numbers w, x, y, z are factors of the product wxyz. However, examination of many whole numbers reveals that some do not have not any factors; while others have two or more factors. For example 11 has no factors, while 84 = 2×2×3×7 has factors 2, 2, 3, 7.

Numbers like 11 are the same as 1×11, however the factor 1 does not add to our knowledge of the number. The factor 1 is referred to as a *trivial* factor. A number which does not have factors, other than 1, is referred to as a *prime number*. For example 2, 3, 5, 7, 11, 13, 17, 19, and 23 are prime numbers. A number which has factors is referred to as a *composite number*. For example, all even numbers greater than 2 are composite numbers, because they have the factor 2.

Prime numbers have received a great deal of attention for a multitude of reasons. If we write down a sequence of prime numbers we get 2, 3, 5, 7, 11, 13, 17, 19, 23, 29, 31, 37, 41, 43, 47, etc. No regular order, or law of formation of this sequence has been found.

When we factor any composite number into its factors we observe that all of the factors are prime numbers. If a factor is not a prime number, then we have not finished the factoring task. Over the years mathematicians have proved the Fundamental Theorem of Arithmetic, which in effect states that *every integer is the product of prime numbers*. Any prime number is simply the product of one prime, itself. Any composite number is the product of two or more primes. For example,

8 *is the same as* $2 \times 2 \times 2$ *i.e.* $8 = 2 \times 2 \times 2$

$120 = 12 \times 10 = 6 \times 2 \times 5 \times 2 = 3 \times 2 \times 2 \times 5 \times 2$

$81 = 9 \times 9 = 3 \times 3 \times 3 \times 3$

$11 = 11 \times 1$

Finding the factors of many numbers is a straightforward, perhaps, tedious process. Finding the factors of some numbers might take a lifetime. Cryptographic codes depend on this fact,

Multiplying The result of a multiplication is called a product. Consider 5 times 3. Here are most of the ways to write a product of two numbers. The parenthesis are required to distinguish the number 53 from the product 5×3.

5*3 *or* 5·3 *or* 5×3 *or* 5(3) *or* (5)3 *or* (5)(3)

Multiplication is defined as a sum of terms.

$5 \times 3 = 3+3+3+3+3$ *all terms are equal to* 3

$3 \times 5 = 5+5+5$ *all terms are equal to* 5

$5 \times 3 = 3 \times 5 = 15$

The minimum possible product of two one digit numbers is 00 and we will show that the maximum possible product of two one digit numbers is 81. Acquire the habit of thinking of each product of two one digit numbers as a two digit number such as 00, 09, 72.

Multiplication uses the carry We can multiply two digit number 26 by one digit number 7 by expanding 26 into 20 + 6. Then the product

$$26 \times 7 = (20+6) \times 7 = (20 \times 7) + (6 \times 7) = 10(2 \times 7) + 1(6 \times 7) = 182$$

```
014    ←carries
 026
×   7
─────
 182
```

The standard algorithm layout is as shown above. The 2 in product 7×6=42 is placed in answer position 0 and the 4 is placed in carry position 1. The 8 from 18=(7×2)+4 is placed in answer position 1, and the 1 is placed in carry position 2. The 1 from 01=(7×0)+1 is placed in answer position 2, and the 0 is placed in carry position 3. Done.

This process is explained in Section 3.3.

3.2 The Multiplication Table, Level 0

The basic tool for doing multiplication is the multiplication table.

The product of any number times zero equals zero. Nine times zero is zero. The product of any number times one equals the number. Five times one equals five. We use elementary algebra to simplify the discussion. Suppose the letter n represents any number. Then we can say

$n \times 0 = 0$ *for any value of n*   *e.g* $7 \times 0 = 0$

$n \times 1 = n$ *for any value of n*   *e.g* $4 \times 1 = 4$

The addition table (page 21) was formed by adding 1 to each number in a row to create the next row. The multiplication table is created in a similar way. Row 0 is all zero's, because $n \times 0 = 0$. Row 1 is 0 to 9, because $n \times 1 = n$. Row 2 is 2 times row 1 so add row 1 to row 1 to create row 2.

Continue by adding row 1 to row n to create row n+1 to produce the complete multiplication table only using addition. We end with row 9 equal to 9 times row 1.

$r\downarrow\ c\rightarrow$	0	1	2	3	4	5	6	7	8	9
0	0	0	0	0	0	0	0	0	0	0
1	0	1	2	3	4	5	6	7	8	9
2	0	2	4	6	8	10	12	14	16	18
3	0	3	6	9	12	15	18	21	24	27
4	0	4	8	12	16	20	24	28	32	36
5	0	5	10	15	20	25	30	35	40	45
6	0	6	12	18	24	30	36	42	48	54
7	0	7	14	21	28	35	42	49	56	63
8	0	8	16	24	32	40	48	56	64	72
9	0	9	18	27	36	45	54	63	72	81

3.3 The Standard Multiplication Algorithm, Level 1

In effect we have multiplied one digit numbers (e.g. 2×7) and produced the multiplication table. At the next level, the multiplication table is used as a tool to multiply a many digit number by a one digit number such as 9361×5. We start by expanding the arbitrarily selected number 9361 as a sum of terms.

$$9361 = 9000 + 300 + 60 + 1$$

We multiply 9361 by the one digit multiplier 5 to produce the sum of the (weighted) products of the digits of 5×9361.

$$5 \times 9361 = 5 \times 9000 + 5 \times 300 + 5 \times 60 + 5 \times 1$$
$$= 45000 \quad + 1500 \quad + 300 \quad + 5 = 46805$$

Observe that expansion relates the 4 digit problem to four one digit problems (5×9, 5×3, 5×6, 5×1). We will show that this is what the standard multiplication algorithm does.

For comparison purposes we relate the addition of 5 copies of 9361 to *non standard* multiplication of 5×9361.

9361	4130	← *carries*
× 5	9361	
05	9361	
300	9361	
1500	9361	
+ 45000	+9361	
46805	46805	

Multiplication is a shorthand notation for addition of quantities (5) of the same number (9361). Motivation to multiply instead of adding intensifies when the multiplier is a large number such as 375.

The multiplication procedure we just executed to multiply 9361 by 5 is OK, just OK, for 1 digit multipliers. It is not useful when the multiplier has two or more digits. This comment is clarified in Section 3.4.

Arithmetic

Standard Multiplication Algorithm, Level 1 Multiply 5 × 9361.

Step 1: Execute a one digit multiplication-with-carry in position 0.

Multiply: $5 \times 1 + 0 = 05 = 0\times10 + 5$
Enter the unit's 5 in answer position 0, and the ten's 0 in carry position 1.

```
      0
   9361
×     5
      5
```

Step 2: Execute a one digit multiplication-with-carry in position 1.

Multiply: $5 \times 6 + 0 = 30 = 3\times10 + 0$
Enter the unit's 0 in answer position 1, and the ten's 3 in carry position 2.

```
     30
   9361
×     5
     05
```

Step 3: Execute a one digit multiplication-with-carry in position 2.

Multiply: $5 \times 3 + 3 = 18 = 1\times10 + 8$
Enter the unit's 8 in answer position 2, and the ten's 1 in carry position 3.

```
    130
   9361
×     5
    805
```

Step 4: Execute a one digit multiplication-with-carry in position 3.

Multiply: $5 \times 9 + 1 = 46 = 4\times10 + 6$
Enter the unit's 6 in answer position 3, and the ten's 4 in carry position 4.

```
   4130
   9361
×     5
   6805
```

Step 5: Execute a one digit multiplication-with-carry in position 4.
Multiply: $5 \times 0 + 4 = 04 = 0\times10 + 4$
Enter the unit's 4 in answer position 4, the ten's 0 in carry position 5. Done

```
  04130
   9361
×     5
  46805
```

The standard multiplication algorithm converted the one 5 digit problem into five 1 digit problems.

```
 9361        9       3       6       1
  ×5    →   ×5      ×5      ×5      ×5
46805       45      15      30      05
```

3.4 The Standard Multiplication Algorithm, Level 2

Multiply 375 × 9361. Show the carries at each step.

Step 1: Execute a level 1 multiplication by 5 in position 0 (Section 3.3).

Multiply: 5 × 9361 = 46805
Enter the result 46805 in the first line below the × line.

```
          4130
          9361
  ×        375
         46805

  +
         _____
```

Step 2: Execute a level 1 multiplication by 7 in position 1.

Multiply: 7× 9361 = 65527
Enter the result 65527 in the second line below the × line.

Note: Digit 7 has value 70 in 375. 65527 is placed one digit left, because its value is really 655270.

```
          6240
          9361
  ×        375
         46805
         65527
  +
         _____
```

Step 3: Execute a level 1 multiplication by 3 in position 2.

Multiply: 3× 9361 = 28083
Enter the result 28083 in the third line below the × line.

Note: Digit 3 has value 300 in 375. 28083 is placed two digits left, because its value is really 2808300.

```
          2110
          9361
  ×        375
         46805
         65527
  + 28083
         _____
```

Last step Add the level 1 results to produce the answer. Addition carries are not shown.

Done.

```
          9361
  ×        375
         46805
         65527
  + 28083
         _____
        3510375
```

Arithmetic

Examples

574 × 38 We calculate the products of 5, 7, and 4 times one digit multipliers 8 and 3. Then we add the position weighted results.

```
 121        453
 574        574      →            574
× 3        ×  8                 × 38
-----      -----                -----
1722       4592                  4592
                               + 1722
                               ------
                                21812
```

The algorithmic process is shown in the layout at the right. The carries are not shown. We keep them in our mind as we proceed.

709 × 123 This example is a four step process. One step for each digit in the multiplier plus one addition step. Observe that the number of zeros added to each partial product equals the multiplier's position number.

```
 000        101        202
 709        709        709                709
× 1       ×   2      ×   3              ×  123
----      ------     ------             ------
                      2127      →         2127
           14180                          1418
70900                                   + 709
                                        ------
                                         87207
```

749 × 403 This is another four step process.

```
 213        000        212
 749        749        749                749
× 4       ×   0      ×   3              ×  403
----      ------     ------             ------
                      2247      →         2247
           00000                          0000
299600                                    2996
                                        ------
                                         301847
```

3.5 Composite and Prime Numbers

Multiplying numbers creates a product. The original numbers are *factors* of the product. Numbers w, x, y, z are factors of the product wxyz. Numbers such as these are *composite* numbers.

$6 = 3 \times 2$

$12 = 2 \times 2 \times 3$

$150 = 5 \times 2 \times 5 \times 3$

Multiplication by 1 does not change the number. So 1 is considered to be a trivial factor.

Observe that there are numbers which do not have factors except the trivial factor 1. They are *prime* numbers.

$2, 3, 5, 7, 11, 13, 17, 19, 23, 29, 31, 37,$ .... *are primes*

You and the world are assured that *all composite numbers are products of primes,* because the Fundamental Theorem of Arithmetic has been proven to be true. The proof is presented when you study Number Theory.

Fundamental Theorem of Arithmetic

> *For each integer* $n > 1$ *there exist primes*
> $p_1 \leq p_2 \leq p_3 \leq \cdots \leq p_r$ *such that*
> $n = p_1 \times p_2 \times p_3 \times \cdots \times p_r$
> *This factorization is unique*

3.6 Review

Multiplicand, Multiplier, Product, Factors   If we want to give 3 books to the 24 members of our class, then we multiply the *multiplicand* 3 by the *multiplier* 24 to find the *product* 72, which is how many books we need. Multiplying 3 by 24 replaces the equivalent addition problem, which is adding 3 twenty-four times. Observe that an easier, equivalent, task is adding 24 three times. The numbers 3 and 24 are *factors* of 72.

multiplicand	× *multiplier*	= *product*
3	× 24	= 72
24	× 3	= 72

Multiplication - The Basic Idea   Expansion of the multiplier converts the original problem into a sum of one digit multiplication problems, each multiplied by their position weight. The one digit problem solutions are found in the Multiplication Table.

The Multiplication Table   Products of two numbers, such as 4 and 7, can be calculated by addition as seven fours 4+4+4+4+4+4+4 or four sevens 7+7+7+7. A less tedious method adds row 1 to build the next row, which is a straightforward way to build the multiplication table. The table shows the 100 products of the one digit numbers 0 to 9. The products range from 00 to 81. We refer to the table's one digit products as level 0.

Products of One digit Multipliers   The procedure for multiplication of two numbers such as 7 and 385 is based on the expansion of the larger of the two numbers. For example, the multiplicand 385 is expanded by digits into a sum of terms 300+80+5. Then each term is multiplied by 7 to get products 2100, 560, and 35. Finally the three products are added to get 2695, which is the product of 7 and 385. The actual process is executed in a more efficient (and obscure) way as follows. We refer to multiplication of any number (385) by a one digit number (7) as level 1.

```
 253
 385
×  7
─────
2695
```

Products of Many Digit Multipliers   A five digit multiplier requires calculating five products of one digit multipliers (level 1) and adding the results. In other words products of many digit numbers are found by executing a sequence of one digit multiplication's (level 1) and adding. We refer to multiplication of any number by any number as level 2.

Multiplying Negative Numbers   There are four possible product outcomes whose sign is as follows. We understand that $- \times - = +$ may be difficult to accept.

$$+x \quad \times \quad +y \quad = \quad +xy$$
$$+x \quad \times \quad -y \quad = \quad -xy$$
$$-x \quad \times \quad +y \quad = \quad -xy$$
$$-x \quad \times \quad -y \quad = \quad +xy$$

Composite and prime numbers   Every integer has factors, which are primes so that the integer equals the product of the prime factors. The primes are the factors of that integer, which is referred to as a composite number. If 1 and p are the only factors of integer p, then p is prime.

$2 \cdot 2 \cdot 3 \cdot 5 = 60$ *60 is a composite number whose factors are* $2, 2, 3, 5$

$1, 2, 3, 5, 7, 11, 13, 17, 19, 23, 29, 31,$ ← *prime numbers*

How the Multiplication Algorithm Works   In effect the algorithm peels off the next position's right hand digits in each step.

$x = 9361 \times 5$

$= 9360 \times 5 + 1 \times 5$	*step*
$= 9360 \times 5 + 5$	1
$= 9300 \times 5 + 60 \times 5 + 5$	
$= 9300 \times 5 + 300 + 5$	2
$= 9000 \times 5 + 300 \times 5 + 300 + 5$	
$= 9000 \times 5 + 1500 + 300 + 5$	
$= 9000 \times 5 + 1000 + 800 + 5$	3
$= 45000 + 1000 + 800 + 5$	
$= 46000 + 800 + 5$	4
$= 46805$	5

4 Subtraction

We start by learning how to *subtract any two one-digit numbers*, such as 9 and 3. Next we learn how to use the *borrow*, and how to apply it to subtraction of numbers with more than one digit. The Standard Subtraction Algorithm implements these ideas.

We learned in *addition* that the symbol + is used when a quantity is increased. That is to say, one number is *added* to another. The + symbol is placed before the second number 1 as in 3 + 1. This is read as three *plus* one. The + symbol can be used to indicate the sum of more than two numbers as in 3 + 1 + 5 + 7.

On the other hand the symbol − is used when a quantity is decreased. That is to say, one number is *subtracted* from another. The − symbol is placed before the second number 1 as in 7 − 1. This is read as seven *minus* one. The − symbol can be used to indicate the subtraction of more than two numbers as in 7 − 1 − 4.

Both symbols can be used in an expression of numbers as in 7 + 3 − 2. Numbers with no symbol, or a +, before them are positive quantities. Numbers with a − before them are *negative* quantities. Use + and − to show a series of increases and decreases. We gather the numbers with a + before them and calculate their sum. Next, we gather the numbers with a − before them and calculate their sum. Then we subtract this sum of negative numbers from the sum of positive numbers. Observe that the result does not change if we form the sums of the numbers in various ways so long as we attend to the symbols + and − before the numbers.

The symbols + and − are referred to as the *signs* of the numbers they are in front of. One example is a calculation of our property. Denote what we own by a + number and what we owe by a − number. If we have $56 and owe one of our friends $23, then our net assets are $56 − $23 or $33. If we own zero dollars and have debts of $23, then our net assets are −$23, which is less than zero. Negative numbers represent less than zero, and positive numbers are greater than zero.

When we add 1 to 0, we have the positive number 1 so that $0 + 1$ is the same as 1. This is the first of the series of positive *natural* numbers. Add 1 without end to create numbers as large as we please. When we subtract 1 from 0, we create the negative number -1 $(0 - 1 = -1)$. This is the first of the series of negative *natural* numbers. Subtract 1 from -1 $(-1 - 1)$ to get -2, and so forth to generate the series of natural numbers $0, -1, -2, -3, ...$ without end to minus infinity.

Observe that all of the following expressions are equal to zero.
$+1 - 1$, $+2 - 2$, $+3 - 3$, $+4 - 4$

Again, if we have \$2 and owe \$7, then we owe \$5. That is to say -5 is the same as $+2 - 7$. The same can be said if we use letters in place of numbers. Consider the expression $+x - y$. There are two cases.

First, if x (27) is greater than y (9), then *subtract y from x* to get z, and place a $+$ before the result z. I.e. $27 - 9 = 16$ $(x-y = z)$, and $z = +16$

Second, if x (5) is less than y (12), then *subtract x from y* to get z, and, this time, place a $-$ before the result z. I.e. $12 - 5 = 7$, and $z = -7$.

These positive and negative numbers are *whole numbers*, or *integers*, which are greater than or less than zero.

Subtraction is the inverse operation to addition. If we subtract 1 from any number whose symbol is n, then the difference is $n-1$ (subtract 1 from 6 to get 5, subtract 1 from 12783 to get 12782, subtract 1 from -6 to get -7). We have said that if we start from 0, then we can create the sequence of numbers from zero to minus infinity by subtracting 1's. This process creates the negative integers $-1, -2, -3, ...$. However we want to do more with subtraction.

We want to know how to subtract any two numbers. Once we know how to do that we can subtract a third number from the result to find the new result. In this way we can subtract any quantity of numbers. So how do we subtract any two numbers? We start with subtraction of one-digit numbers, and acquire an understanding of the borrow idea, which is the inverse of the carry idea. Then we use the Standard Subtraction Algorithm to subtract any two numbers with any numbers of digits.

4.1 Negative Numbers

The logical desire for the subtraction operation forced the creation of negative numbers so that subtraction is possible in any problem. The expression 32–58 has no meaning in the domain of positive integers. When we write 32–58 = –26 the negative number –26 is created by subtraction.

The ideas of subtraction and negative numbers are fairly easily illustrated on a *number line*. Number line is a geometrical representation of number. The positive side of the number line was presented in Section 1.1. We now show the number line has a negative side.

Review: The number line is constructed by marking off equal distances along the line. Each mark on the number line is assigned a number. Assign 0 to any mark. Next assign 1 to the first mark to the *right* of zero. Then the *distance* from 0 to 1 *represents* 1 unit of length. Subsequent marks to the right add 1 unit to the distance. This means subsequent marks represent 2, 3, 4, and so forth. In other words numbers increase by one at each successive mark to the right of zero.

Pretend we have a ladder with many rungs, which we have placed in a hole so that one rung is at ground level. Label that rung zero (0). Furthermore, pretend the rungs above ground correspond to the positive side of the number line. If we climb up the ladder we step on rungs 0, 1, 2, etc. When we climb back down we return to rungs 2, 1, 0. If we continue climbing down we go below ground, and we say we step on rungs –1, –2, –3, etc. What does this mean? Well, +/– means above/below ground. How about a temperature scale where +/– means above/below the freezing temperature zero degrees Centigrade?

```
-8 -7 -6 -5 -4 -3 -2 -1  0  1  2  3  4  5  7  8  9
└──┴──┴──┴──┴──┴──┴──┴──┴──┴──┴──┴──┴──┴──┴──┴──┘
```

The key to using the number line is to count spaces between the marks. The *distance* from 0 to 1 *represents* 1. The distance from 1 to 2 represents 1 more. Watch this. The distance from –3 to –2 also represents 1 more. The distance from 2 to 1 represents 1 less. The distance from –1 to –2 also represents 1 less. Move to the right to increase value. Move to the left to decrease value.

Subtraction creates a new type of number, the negative integer when the number of objects removed is greater than the number of objects available. This is not absurd, because in the abstract this is possible.

> In many elementary arithmetic curriculums negative numbers are *avoided* by never presenting a problem where the number of objects removed is greater than the number of objects available. Omission of negative numbers leaves a logical gap in the minds of thinking people. This is why we do not avoid negative numbers.

Nevertheless there is a way people use so that the number of objects removed is always smaller than the number of objects available. The way is to exchange the numbers and mark the result as negative. We do *not* perform that artificial maneuver when we are in the process of solving a complex problem. Actually we never even think of performing it.

Pretend we are standing on mark zero. Seven steps to the right puts us on mark 7. Each step covers a distance of one. Then four steps to the left returns us to mark 3. We can interpret seven steps right as +7 and four steps left as –4 (take away 4), so that $7 - 4 = 3$.

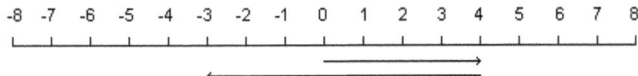

The numbers 7 and 4 are positive. We placed a minus sign before the 4 to indicate subtraction of 4 from 7, because the steps were 7 to the right and 4 to the left. Since 7 is greater than 4 the result is the positive number 3. Suppose the numbers 7 and 4 are exchanged.

Start from 0, walk right 4 steps, and then walk left 7 steps to end on the –3 mark (count up, then down: right 1, 2, 3, 4, now left 3, 2, 1, 0, –1, –2, –3 stop). The net change or difference $d = 4 - 7 = -3$.

We know the difference is 3. The minus sign means we took more steps left than we took to the right. So we end up 3 steps left of zero. When we walked 7 steps left we passed 0 (the start position). We say d = 4 – 7 = –3 (negative 3). We also say d is less than zero.

The symbol for *less than* is <.

The expression d < 0 means *d is less than zero.*

The symbol for *greater than* is >.

The expression d > 0 means *d is greater than zero.*

This is why we marked numbers to the left of zero with a minus sign.

For many students the first encounter with negative numbers is fierce; what is this thing called negative?

A negative number is simply a quantity less than zero.

That is all there is to it. A negative number is less than zero.

Think relatively. If you have 23 pennies and I have 12, then I have 11 less than you have, or I could say my wealth is –11 with respect to yours. Do you have any difficulty saying the temperature is ten degrees below zero, or minus ten degrees?

How about saying + and – are inverses; i.e. opposites? Why not? After some thought we will conclude correctly that positive integers are positive *by definition*; so we are consistent when we *define* negative integers as opposites of positive integers

4.2 The Standard Subtraction Algorithm

Subtract 56913 from 67958 using the Standard Subtraction Algorithm, which uses the equations defining the numbers. Observe that all steps are identical, and that they use the addition table. Section 4.3 considers column sums that are negative, which require borrowing.

Step 1: Execute a one digit subtraction in position 0.

Subtract: units 8–3 = 5 (what do we add to 3 to produce 8?)
Enter the difference 5 in answer position 0.

$$\begin{array}{r} 67958 \\ -\,56913 \\ \hline 5 \end{array}$$

Step 2: Execute a one digit subtraction in position 1.

Subtract tens 5–1 = 4 (what do we add to 1 to produce 5?)
Enter the difference 4 in answer position 1.

$$\begin{array}{r} 67958 \\ -\,56913 \\ \hline 45 \end{array}$$

Step 3: Execute a one digit subtraction in position 2.

Subtract: hundreds 9–9 = 0 (what do we add to 9 to produce 9?)
Enter the difference 0 in answer position 3.

$$\begin{array}{r} 67958 \\ -\,56913 \\ \hline 045 \end{array}$$

Step 4: Execute a one digit subtraction in position 3.

Subtract: thousands 7–6 = 1 (what do we add to 6 to produce 7?). Enter the difference 1 in answer position 3.

$$\begin{array}{r} 67958 \\ -\,56913 \\ \hline 1045 \end{array}$$

Step 5: Execute a one digit subtraction in position 4.

Subtract: ten thousands 6–5 = 1 (what do we add to 5 to produce 6?). Enter the difference 1 in answer position 4.

$$\begin{array}{r} 67958 \\ -\,56913 \\ \hline 11045 \end{array}$$

We can check this result by expanding each number as a sum of terms and subtracting term by term.

$67958 = 60000 + 7000 + 900 + 50 + 8$
$56913 = 50000 + 6000 + 900 + 10 + 3$
the difference is
$11045 = 10000 + 1000 + 000 + 40 + 5$

Arithmetic

The standard algorithm layout for 67958–56913, and five one digit equations used in the algorithm are

67958
–56913   →   6 – 5 = 1   7 – 6 = 1   9 – 9 = 0   5 – 1 = 4   8 – 3 = 5
11045

The digits in any one column have the same weight. I.e. there are five 10's in 67958 and one 10 in 56913. So there are four 10's in the difference.

The subtraction algorithm starts at the right by subtracting the 1's digits, moves one position left and subtracts the 10's digits, and so forth, to produce the column differences 5, 4, 0, 1, 1 as we use the Addition Table.

We can rewrite the five one digit equations to emphasize position and position weight.

60000	7000	900	50	8
– 50000	– 6000	– 900	– 10	– 3
10000	1000	000	40	5

The subtraction of 5 digit numbers is converted to five subtractions of one digit numbers. A conversion facilitated by the ideas of position and position weight. In other words the subtraction of each column of digits is *independent* of the other columns.

4.3 The Standard Subtraction Algorithm and the Borrow

Subtraction of two one digit numbers produces positive or negative one digit differences ranging from 9 to –9.

9	6	5	3	0
-0	-2	-3	-7	-9
9	4	2	-4	-9

The positive results present no problems when subtracting one digit numbers. The negative results violate the *all results positive rule*. Here is the key idea we use when doing subtraction.

> Adding 10 *always produces* a positive result, because $10 - 9 = 1 > 0$.

Borrowing Borrowing implements the key idea. Take (subtract) 1 ten from a column (position n+1) and mark that event by adding –1 to the column. Exchange one ten for ten ones. Add the ten ones to the column at the right (position n). The value of the result will not change, because –1 ten plus 10 ones equals 0. Taking a ten from the column at the left is called *borrowing*.

> *A borrow converts 1 ten into 10 ones. This is the*
> *inverse of a carry, which converts 10 ones to 1 ten.*

In this way a ten is converted into ones thereby increasing the number of ones by ten. This guarantees that a positive number is the result of the subtraction in any column. A positive result is guaranteed because 10 is greater than any one-digit number 0 to 9. This is the *borrow idea*.

For example, subtract 6 from various two digit numbers to apply the borrow idea. Verify the answer by adding the answer to the number subtracted. E.g. 26+6=32.

0	10	10	10	
0	-1	-1	-1	← *borrows*
9 7	1 4	3 2	1 0	
$-$ 6	$-$ 6	$-$ 6	$-$ 9	
9 1	0 8	2 6	0 1	

Arithmetic

Subtract 56913 from 76381 using the Standard Subtraction Algorithm.

Step 1: Execute a one digit subtraction-with-borrow in position 0.

0+1−3=−2 Must borrow.
Enter −1 in borrow position 1 and 10 in borrow position
0. Algebraic sum: 10+1−3 = 8
Enter the 8 in answer position 0.

				10
			−1	
7	6	3	8	1
−5	6	9	1	3
				8

Step 2: Execute a one digit subtraction-with-borrow in position 1.

−1+8−1=6 No borrow.
Algebraic sum: −1+8−1 = 6
Enter the 6 in answer position 1.

				10
			−1	
7	6	3	8	1
−5	6	9	1	3
			6	8

Step 3: Execute a one digit subtraction-with-borrow in position 2.

3−9=−6 Must borrow.
Enter −1 in borrow position 3 and 10 in borrow position
2. Algebraic sum: 10+3−9 = 4
Enter the 4 in answer position 2.

		10		10
		−1		−1
7	6	3	8	1
−5	6	9	1	3
		4	6	8

Step 4: Execute a one digit subtraction-with-borrow in position 3.

−1+6−6=−1 Must borrow.
Enter −1 in borrow position 4 and 10 in borrow position
3. Algebraic sum: 10−1+6−6 = 9
Enter the 9 in answer position 3.

	10	10		10
	−1	−1		−1
7	6	3	8	1
−5	6	9	1	3
	9	4	6	8

Step 5: Execute a one digit subtraction-with-borrow in position 4.

−1+7−5=1 No borrow.
Algebraic sum: −1+7−5 = 1
Enter the 1 in answer position 4. Done.

	10	10		10
	−1	−1		−1
7	6	3	8	1
−5	6	9	1	3
1	9	4	6	8

4.4 Examples

A major consequence of the borrow idea is that the borrow allows us to subtract the digits in any column *independently* of the digits in other columns. (The carry did the same for the addition process.) The 5 digit problem is solved one digit at a time.

Emphasis The mathematics community says we borrow 1 ten from column n+1, mark it as −1, and we enter the 10 at the top of column n to be added later when we sum column n. This is what the Standard Subtraction Algorithm does.

> *The borrow allows us to satisfy the condition that*
> *only positive results are allowed in any position, and*
> *to subtract each column independently.*

> *Borrowing while subtracting replaces carrying while adding.*

Examples with base 10

100 − 99 In the ones column 0 minus 9 equals −9 so a borrow is required. We ignore the 0 in the tens column and borrow anyway, because we see the 1 to the left. The column of ones sum is $10 + 0 − 9 = 1$.

The new sum of the tens column is $−1−9 = −10$. We must borrow from the hundreds column. Now the column of tens sum is $10 − 1 − 9 = 0$. In the hundreds column the sum is $−1 + 1 − 0 = 0$. Done.

	10	10
−1	−1	
1	0	0
−	9	9
0	0	1

225 − 83 Since 3 is less than 5 a borrow is not required, however in the tens column, 8 is greater than 2 so we borrow ten tens from the 100's column, and mark the borrow by entering −1 above the 100's column. The −1 means take away 100, because the −1 is in the 100's column.

```
        10        ← ten tens added to tens column
  -1              ← ten tens borrowed from 100s column
   2    2    5
  -0    8    3
  ___  ___  ___
   1    4    2
```

51

The ten tens are then added to the tens column. This procedure leaves the value of the result unchanged and guarantees positive results in each column. *Now we can add each column independently*: $5 - 3 = 2$, $10 + 2 - 8 = 4$ and $-1 + 2 - 0 = 1$.

641 – 578 This problem shows how more than one consecutive borrow works. Since 1 minus 8 is minus 7 we must borrow one from the 10's column and add ten to the 1's column. We can borrow, because there are 4 tens. The sum in the ones column is $10 + 1 - 8 = 3$. In the tens column $-1 + 4 - 7 = -4$. We must borrow one from the 100's column and add ten to the 1's column to get 6. In the hundreds column $-1 + 6 - 5 = 0$. Done.

	10	10	*add* 10		
	−1	−1	*borrow* 1	$10 + 1 - 8 = 3$	(*position* 0)
641 →	6	4	1	$10 - 1 + 4 - 7 = 6$	(*position* 1)
− 578	−5	−7	−8	$-1 + 6 - 5 = 0$	(*position* 2)
63	0	6	3		

3085 – 837 Digits with the same weight are in the same position in each number. We use the Addition Table to find the differences 3, −8, 5, and −2 of digits in each column.

Position →	3	2	1	0
3085	3	0	8	5
837	0	8	3	7
Column differences	3	−8	5	−2

Starting at position 0 we must borrow from the left to convert the negative differences into positive one digit numbers, *because only positive one digit numbers can reside in any position*. A borrow leaves a −1 in the 'ten times' column and a 10 in the 'one times' column.

		−1	10	*borrow* 1, *add* 10
	−1	10		*borrow* 1, *add* 10
3085 →	3	0	8	5
− 837	−0	−8	−3	−7
2248	2	2	4	8

Examples with bases not equal to ten These examples *show the power of knowing the theory.* Borrowing subtracts and adds equivalents such as 1 hour and 60 minutes. This is not different in principle from borrowing 1 ten and adding 10 ones.

Base 60 What is elapsed time from time **2:33:49** *to time* **13:18:34?** The relationship between hours and minutes is 60:1. The relationship between minutes and seconds is 60:1. Observe how we use these relationships when we borrow.

hour	minute	second	
	-1	60	*borrow* 1 *minute* = 60 *seconds*
-1	60		*borrow* 1 *hour* = 60 *minutes*
13	18	34	
-02	-33	-49	
10	44	45	

How many days from **09/18/1999** *to* **06/07/2001?** The relationship between days and months is say 30:1. The relationship between months and years is 12:1.

year	month	day	
	-1	30	*borrow* 1 *month* = 30 *days*
-1	12		*borrow* 1 *year* = 12 *months*
2001	6	7	
-1999	-9	-18	
1	8	19	

This example *shows the power of knowing the theory of addition.* We add to verify the result. For example, observe below that the sum of the month column is 18 months, which converts to 1 year and 6 months so that the sum is 6 and the carry is 1.

year	month	day	
	1		*carry* 1 *month* = 30 *days*
1			*carry* 1 *year* = 12 *months*
1999	9	18	$37 - 30 = 7$
$+1$	$+8$	$+19$	$18 - 12 = 6$
2001	6	07	

4.5 Review

Subtraction is done the way it is done, because we can only subtract *digits with the same weight*, and we must *borrow* because only *positive* one-digit numbers can reside in any answer position.

Number Line Number line uses graphics to display the meaning of numbers. The rules are straightforward.

Numbers increase in value when we move to the right along the line. Numbers decrease in value when we move to the left along the line.

```
-8  -7  -6  -5  -4  -3  -2  -1   0   1   2   3   4   5   6   7   8
L___I___I___I___I___I___I___I___I___I___I___I___I___I___I___I___J
```

Negative Numbers Subtraction creates a new type of number, the negative number, when the number subtracted is greater than the number subtracted from. Here is $4-7 = -3$.

```
-8  -7  -6  -5  -4  -3  -2  -1   0   1   2   3   4   5   6   7   8
L___I___I___I___I___I___I___I___I___I___I___I___I___I___I___I___J
                            ←_____→
```

A negative number y is a quantity less than zero ($y<0$).

Borrow Idea A borrow has to be generated when the algebraic sum of any column is less than 0 (negative). The borrow idea converts a one in the column to the left of the current column to ten ones in the current column. This increases the numbers of ones in the current column by 10 to guarantee that a positive number is the result of the subtraction in the current column. Always positive, because 10 is greater than any one-digit number 0 to 9.

Verify a subtraction To verify the result z in $x-y = z$, form the sum y+z. Does it equal x?

5 Division

Division is an efficient way to subtract the same number many times. Addition, subtraction and multiplication are prerequisites for division. The new items to learn are the division process producing quotient and remainder, and the Standard Division Algorithm.

The number to be divided (4521) is the *dividend*. The *divisor* is the number (23) to be subtracted 196 times from the dividend. The *quotient* (196) is how many divisors (23) are in the dividend (4521). The partial divisor (13) left over is the *remainder*.

$$\frac{dividend}{divisor} = quotient + \frac{remainder}{divisor} \;\rightarrow\; \frac{x}{y} = q + \frac{r}{y} \;\rightarrow\; \frac{4521}{23} = 196 + \frac{13}{23}$$

The quotient (q=196) is the largest quantity of the divisor (y=23) that can be subtracted from the dividend (x=4521). The remainder (r=13) is what is left over. The inference is that q+1=196+1=197 multiples of the divisor 23 is greater than 4521. In mathematics these words are represented by a double inequality where < means less than.

$$(196 \times 23) < 4521 < (197 \times 23) \quad means \quad (196 \times 23) < 4521 \;\; and \;\; 4521 < (197 \times 23)$$
$$qy < x < (q+1)y \qquad means \qquad qy < x \quad and \quad x < (q+1)y$$
$$Then \;\; r = x - qy \; < \; (q+1)y - qy = y \quad (remainder \; r \; is \; less \; than \; divisor \; y)$$

---

remainder r ranges from 0 to y–1: $0 \leq r < y$ (y is the divisor)

Dividing 4521 by 23 produces quotient 196 and remainder 13. Remainder 13 tells us 23 is not a factor of 4521. However the divisor 23 is a factor of 4508, because 4508 is dividend 4521 minus remainder 13. The world says 4508 is *divisible* by 23. The factors of 4508 are 196 and 23. Some of the division symbols are the long division symbol, division sign ÷, bar —, and slash /.

$$23\overline{)4521} \qquad 4521 \div 23 \qquad \frac{4521}{23} \qquad 4521/23$$

What we have done so far does not do what a standard algorithm does, which is to convert one N-digit problem into N one-digit problems.

5.1 Prelude to the Standard Division Algorithm

Emphasis Divide 33 by 7. The quotient q is the quantity of divisors y that can be subtracted from the dividend x. (The quotient 4 is the quantity of 7's that can be subtracted from 33.) What remains is a partial divisor y that is referred to as the remainder r=5 that is always less than a whole y, because q of the whole y's have been removed from x.

$$\frac{dividend}{divisor} = quotient + \frac{remainder}{divisor} \quad \rightarrow \quad \frac{x}{y} = q + \frac{r}{y} \quad \rightarrow \quad x = qy + r \quad (33 = 7 \times 4 + 5)$$

Division on the number line Enter marks on the number line that are a multiple of the divisor y. Mark the number x to be divided. Observe that length x is the sum of length qy plus length r. Observe that mark x falls between marks qy and (q+1)y. Mark x falls on qy when the remainder is zero. Clearly remainder r is less than divisor y.

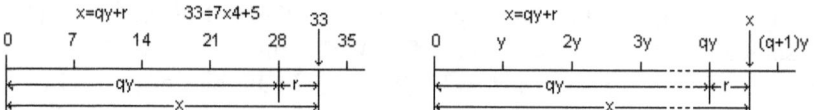

Example Consider 1254 divided by 7. The purpose of division by 7 is to find out how many 7's can be subtracted from 1254. The division process subtracts groups of 7's from 1254 until no whole 7's remain. A partial 7 ranging from 0 to 6 may be left over. This partial seven is the remainder.

An inefficient method subtracts 7's one at a time. In fact the inefficient method subtracts seven 179 times leaving a remainder of 1.

We intend to show that division by 7 is an efficient method for subtracting all of the 7's contained in 1254. An efficient method subtracts groups of 7's at a time, such as 1000, 100, 10. 1. We are proposing for the 1254/7 problem to subtract values of 7000, 700, 70 and 7 (which are multiples of 7) corresponding to positions 3, 2, 1, and 0.

Use the level 0 multiplication table (3.2) to ask what is greatest number of 7's less than the current remainder produced by each step in the division process. Do this in each step. This is the bridge to division.

> *Crucial assumption* If an object is divided into n parts, then all of the parts are of **equal** size (length, volume, whatever).

Step 1 The 1 in 1254 is in position 3 with weight 1000. Since 1000 is less than 7000 we cannot subtract 7000 from 1000. Dividing 7 into the 1 produces quotient 0 and remainder 1.

```
      0          put quotient digit 0 in position 3
  7)1254
   -0            0×7 = 0
   ————
   1254          7000 > (r = 1254) > 7,  continue
```

Step 2 The 2 in 1254 is in position 2 with weight 100 so we ask how many 700's can we subtract from r=1254? The answer is one. Dividing 7 into the 12 produces quotient 1 and remainder 5.

```
     01          put quotient digit 1 in position 2
  7)1254
  -0
  ————
   12            bring down the 2
   -7            1×7 = 7    subtract 100 sevens (value 700)
   ————
   554           700 > (r = 554) > 7, continue
```

Step 3 The 5 in 1254 is in position 1 with weight 10 so we ask how many 70's can we subtract from r=554? The answer is seven. Dividing 7 into the 55 produces quotient 7 and remainder 6.

```
    017          put quotient digit 7 in position 1
  7)1254
  -0
  ————
   12
   -7
  ————
   55            bring down the 5
   49            7×7 = 49    subtract 70 sevens (value 490)
   ————
   64            70 > (r = 64) > 7, continue
```

Step 4 Show that this step divides 64 by 7 with remainder 1.

5.2 The Standard Division Algorithm

The arithmetic algorithms solve any N digit problem one digit at a time. This is exactly what the standard division algorithm does. Division of 1254 by 7 is a N=4 digit problem that is solved by performing 4 one digit division problems which we can do in our mind. We showed in 5.1 that

$$7\overline{)1254} \text{ is converted into } \quad 7\overline{)1} \quad 7\overline{)12} \quad 7\overline{)55} \quad 7\overline{)64}$$

The genius of this algorithm will become clear as we proceed. First an observation on format.

$$\frac{dividend}{divisor} = quotient + \frac{remainder}{divisor} \qquad \frac{m}{d} = q + \frac{r}{d} \quad and \quad m = (q \times d) + r$$

Reminder: always ask how many 7's can we subtract in each step.

1254/7 Dividend$_1$ is 1254. The divisor is 7 in every step. Use the equation $m=(q \times d)+r$ in every step. The letters in the equation represent *dividend=(quotient×divisor)+remainder. The greatest number of 7's that can be subtracted is q sevens.*

Step 1 is different. Steps from 2 to the end are identical procedures.

Step 1: Execute a one digit division where the dividend is the digit 1 of dividend$_1$ = 1254.

Divide 1 by 7. $1 = (0 \times 7) + 1$ $m=(q \times d)+r$
Enter the quotient digit 0 over the leftmost digit 1, subtract 0×7=0 leaving remainder r=1.

```
   0
7)1254
  -0
---
   1
```

Step 2: Multiply the remainder 1 of the preceding step by 10 and add to it the next digit 2 in 1254 to produce dividend$_2$.

Dividend$_2$ = (1 × 10) + 2 = 12.
Divide 12 by 7.   12 =7+5 = (1 × 7) + 5
Enter the quotient digit 1 over the next digit 2, subtract 1×7=7 leaving remainder 5.

```
  01
7)1254
  -0
---
  12
  -7
---
   5
```

Step 3: Multiply the remainder 5 of the preceding step by 10 and add to it the next digit 5 in 1254 to produce dividend$_3$.

Dividend$_3$ = $(5 \times 10) + 5 = 55$.
Divide 55 by 7. $55 = 49+6 = (7 \times 7) + 6$
Enter the quotient digit 7 over the next digit 5, subtract 7×7=49 leaving remainder 6.

```
    017
7)1254
  -0
 ────
   12
  -7
 ────
   55
  -49
 ────
    6
```

Step 4: Multiply the remainder 6 of the preceding step by 10 and add to it the next digit 4 in 1254 to produce dividend$_4$.

Dividend$_4$ = $(6 \times 10) + 4 = 64$.
Divide 64 by 7. $64 = 63+1 = (9 \times 7) + 1$
Enter the quotient digit 9 over the next digit 4, subtract 9×7=63 leaving remainder 1.

Since remainder 1 is less than 7 we are done.

And so 1254/7 is represented by the four one digit division expressions:

```
   0179
7)1254
  -0
 ────
   12
  -7
 ────
   55
  -49
 ────
   64
  -63
 ────
    1
```

$01 = (0 \times 7) + 1$
$12 = (1 \times 7) + 5$
$55 = (7 \times 7) + 6$
$64 = (9 \times 7) + 1$
Verify: $(179 \times 7) + 1 = 1254$

Observe that the quotient 0179 can be perceived by reading vertically down the first column of digits on the right side of =. The final remainder, 1, is the last term of the last equation. The remainder is produced by a series of subtractions.

$$r = 1254 = 1254 - (0 \times 7000) = 1254 \qquad step\,1$$
$$= 1254 = 1254 - (0 \times 7000) - (1 \times 700) = 554 \quad step\,2$$
$$= 1254 = 1254 - (0 \times 7000) - (1 \times 700) - (7 \times 70) = 64 \quad step\,3$$
$$= 1254 = 1254 - (0 \times 7000) - (1 \times 700) - (9 \times 7) = 1 \quad step\,4$$

$$1254/7 = 179 + (1/7)$$

Arithmetic

With experience Here is how we will probably divide when we have mastered the algorithm.

"Bring down the digit" replaces "Multiply the remainder r of the preceding step by 10 and add to it the next digit n in dividend$_1$ to produce dividend$_m$."

E.g. "7 goes into 55 7 times" replaces "divide 55 by 7 to get 7."

This is how practical procedures obscure the algorithm.

```
    0179
7)1254                          Divide 1 by 7
  -0
 ────
   12      Bring down the 2. This is equivalent to 1×10+2. Divide 12 by 7
  -7
 ────
   55      Bring down the 5. This is equivalent to 5×10+5. Divide 55 by 7
 -49
 ────
   64      Bring down the 4. This is equivalent to 6×10+4. Divide 64 by 7
 -63
 ────
    1                           0 < remainder 1 < 7, Done.
```

Division is not a guessing game In step 3 we are asked to divide 64 by 7. In our mind we are supposed to draw upon our knowledge of the level 0 multiplication table (Section 3.2) to determine the largest number of 7's contained in 64. Our mental calculations may proceed as follows.

Eight 7's equal 56 with remainder 8, which is greater than 7. Nine 7's equal 63 with remainder 1, which is less than 7. Ten 7's (that's a no-no in any event) equal 70 with remainder −6. Clearly the correct answer is nine 7's where $64 = (9 \times 7) + 1$. And so 9 is the next digit in the quotient.

Bad advice In many classes students are advised (taught?) *to guess* at the next quotient digit. There is no need to guess if we know the level 0 multiplication table (page 34).

5.3 Division by Numbers with any number of digits

Addition and multiplication algorithms start with creation of tables involving one digit numbers 0 to 9, and build from there. There are no division tables to build upon. The standard division algorithm requires us to do division problems such as 45/23 in our mind, which arises from the digits 45 in the dividend 45224. Use subtraction, e.g. 45–23, and the multiplication table to solve these elementary problems in your mind. The divisor 23 has two digits. The standard algorithm proceeds in the same way when the divisor has any number of digits. Here is 45224/23 executed by the division algorithm.

45224/23 Dividend$_1$ is 45224. The divisor is 23. *Observe that the greatest number of 23's is the q in the equation $m=(q\times d)+r$, which means dividend=(quotient$\times$divisor)+remainder.*

Step 1: Execute a one digit division where the dividend is the leftmost digit 4 of dividend$_1$ = 45224.

Divide 4 by 23. $4 = (0 \times 23) + 4$ $m=(q \times d)+r$
Enter the quotient digit 0 over the leftmost digit 4, subtract 0×23=0 leaving remainder r=4.

0
23⟌45224
−0
4

Step 2: Multiply the remainder 4 of the preceding step by 10 and add to it the next digit 5 in 45224 to produce dividend$_2$.

Dividend$_2$ = $(4 \times 10) + 5 = 45$.
Divide 45 by 23 $45 = (1 \times 23) + 22$
Enter the quotient digit 1 over the next digit 5, subtract 1×23=23 leaving remainder 22.

01
23⟌45224
−0
45
−23
22

Step 3: Multiply the remainder 22 of the preceding step by 10 and add to it the next digit 2 in 45224 to produce dividend$_3$.

Dividend$_3$ = $(22 \times 10) + 2 = 222$.
Divide 222 by 23 $222 = (9 \times 23) + 15$
Enter the quotient digit 9 over the next digit 2, subtract 9×23=207 leaving remainder 15.

019
23⟌45224
−0
45
−23
222
−207
15

Arithmetic

Step 4: Multiply the remainder 15 of the preceding step by 10 and add to it the next digit 2 in 45224 to produce dividend$_4$.

Dividend$_4$ = (15 × 10) + 2 = 152.
Divide 152 by 23   152 = (6 × 23) + 14
Enter the quotient digit 6 over the next digit 2, subtract 6×23=138 leaving remainder 14.

```
      0196
23) 45224
    -0
     45
    -23
     222
    -207
      152
     -138
       14
```

Step 5: Multiply the remainder 14 of the preceding step by 10 and add to it the next digit 4 in 45224 to produce dividend$_5$.

Dividend$_5$ = (14 × 10) + 4 = 144.
Divide 144 by 23   144 = (6 × 23) + 6
Enter the quotient digit 6 over the next digit 4, subtract 6×23=138 leaving remainder 6.

Since 6 is less than 23 we are done.

```
     01966
23) 45224
    -0
     45
    -23
     222
    -207
      152
     -138
       144
      -138
         6
```

Recapitulation
004 = (0 × 23) + 4
045 = (1 × 23) + 22
222 = (9 × 23) + 15
152 = (6 × 23) + 14
144 = (6 × 23) + 6

Answer $45224 = 1966 \times 23 + 6$

Observe that the quotient 01966 can be perceived by reading vertically down the first column of digits on the right side of equals. The final remainder is 6 (the last term of the last equation).

Divide 100000 by 301 Dividend$_1$ is 100000. The divisor is 301 in every step. Use the equation $m=(q \times n)+r$ in every step.

Step 1: Execute a one digit division where the dividend is the leftmost digit 1 of dividend$_1$ = 100000.

```
        0
301) 100000
    - 0
      1
```

Divide 1 by 301. $1 = (0 \times 301) + 1$
Enter the quotient digit 0 over the leftmost digit 1, subtract 0×301=0 leaving remainder r=1.

Step 2: Multiply the remainder 1 of the preceding step by 10 and add to it the next digit 0 in dividend$_1$ to produce dividend$_2$.

```
       00
301) 100000
    - 0
      10
     - 0
      10
```

Dividend$_2$ = $(1 \times 10) + 0 = 10$.
Divide 10 by 301. $10 = (0 \times 301) + 10$
Enter the quotient digit 0 over the next digit 0, subtract 0×301=0 leaving remainder r=10.

Step 3: Multiply the remainder 10 of the preceding step by 10 and add to it the next digit 0 in dividend$_1$ to produce dividend$_3$.

```
      000
301) 100000
    - 0
      10
     - 0
      100
     - 0
      100
```

Dividend$_3$ = $(10 \times 10) + 0 = 100$.
Divide 100 by 301. $100 = (0 \times 301) + 100$
Enter the quotient digit 0 over the next digit 0, subtract 0×301=0 leaving remainder r=100.

Step 4: Multiply the remainder 100 of the preceding step by 10 and add to it the next digit 0 in dividend$_1$ to produce dividend$_4$.

```
     0003
301) 100000
    - 0
      10
     - 0
      100
     - 0
      1000
     - 903
       97
```

Dividend$_4$ = $(100 \times 10) + 0 = 1000$.
Divide 1000 by 301. $1000 = (3 \times 301) + 97$
Enter the quotient digit 3 over the next digit 0, subtract 3×301=903 leaving remainder r=97.

Step 5: Multiply the remainder 97 of the preceding step by 10 and add to it the next digit 0 in dividend$_1$ to produce dividend$_5$.

Dividend$_5$ = (97 × 10) + 0 = 970.
Divide 970 by 301.   970 = (3 × 301) + 67
Enter the quotient digit 3 over the next digit 0, subtract 3×301=903 leaving remainder r=67.

```
      00033
301) 100000
    - 0
    ─────
      10
    -  0
    ─────
      100
    -   0
    ─────
      1000
    -  903
    ─────
        970
      - 903
      ─────
         67
```

Step 6: Multiply the remainder 67 of the preceding step by 10 and add to it the next digit 0 in dividend$_1$ to produce dividend$_6$.

Dividend$_6$ = (67 × 10) + 0 = 670.
Divide 670 by 301.   670 = (2 × 301) + 68
Enter the quotient digit 2 over the next digit 0, subtract 2×301=602 leaving remainder r=68.

0001 = (0 × 301) + 1
0010 = (0 × 301) + 5
0100 = (0 × 301) + 100
1000 = (3 × 301) + 97
0970 = (3 × 301) + 67
0670 = (2 × 301) + 68
Verify: (332 × 301) + 68 = 100000

```
     000332
301) 100000
    - 0
    ─────
      10
    -  0
    ─────
      100
    -   0
    ─────
      1000
    -  903
    ─────
        970
      - 903
      ─────
        670
      -  602
      ─────
          68
```

5.4 Division by zero has no meaning

To divide x by y means to find a number q such that

$x = y \times q$

If $y = 0$, then $x = 0 \times q = 0$ for ANY finite value of q.

Observe that ANY number for q is a solution to $0 = 0 \times q$

Therefore $\dfrac{x}{0}$ is an indeterminate form that has no meaning

5.5 Dividing with Negative Numbers

So far all numbers have been positive. However, the dividend's sign can be either + plus or − minus. The same applies to the divisor. Consequently there are four cases.

If y, z, q, r are positive numbers, then

$$\frac{+y}{+z} = +q + \frac{+r}{+z} \quad because \quad +y = zq + r$$

$$\frac{+y}{-z} = -q + \frac{+r}{-z} \quad because \quad +y = zq + r$$

$$\frac{-y}{+z} = -q + \frac{-r}{+z} \quad because \quad -y = -zq - r$$

$$\frac{-y}{-z} = +q + \frac{-r}{-z} \quad because \quad -y = -zq - r$$

Division has the same rules for signs as multiplication. In practice pretend the numbers y and z are positive, divide, and then assign signs as shown above.

Again - Negative numbers

$$\frac{+ab}{+a} = +b \quad because \quad (+a) \times (+b) \rightarrow +ab$$

$$\frac{+ab}{-a} = -b \quad because \quad (-a) \times (-b) \rightarrow +ab$$

$$\frac{-ab}{+a} = -b \quad because \quad (+a) \times (-b) \rightarrow -ab$$

$$\frac{-ab}{-a} = +b \quad because \quad (-a) \times (+b) \rightarrow -ab$$

5.6 Review

Division is an efficient way to subtract the same number many times.

The number to be divided (4521) is the *dividend*. The *divisor* is the number (23) to be subtracted 196 times from the dividend. The *quotient* (196) is how many divisors (23) are in the dividend (4521). The partial divisor (13) left over is the *remainder*.

$$\frac{dividend}{divisor} = quotient + \frac{remainder}{divisor}$$

$$\frac{4521}{23} = 196 + \frac{13}{23} \quad and \quad 4521 = 196 \times 23 + 13$$

Consider dividing 4521 by 23. The purpose of division by 23 is to find out how many 23's add up to 4521. This is why the division process subtracts 23's from 4521 until no whole 23's remain. A partial 23 ranging from 0 to 22 may be left over. This partial 23 is called the remainder.

Division by 23 is an efficient method for subtracting all of the 23's contained in 4521. An inefficient method subtracts 23's one at a time. In fact the inefficient method subtracts twenty-three 196 times from 4521 leaving a remainder of 13 ($4521 = 23 \times 196 + 13$).

Division by zero has no meaning.

The Standard Division Algorithm converts one N-digit problem into N one-digit problems. The N = 5 digit problem 45224/23 converts into 5 "one digit" problems 4/23, 45/23, 222/23, 152/23, and 144/23 that we can do in our mind.

Division on the number line Enter marks on the number line that are a multiple of the divisor y. Mark the number x to be divided. Observe that length x is the sum of length qy plus length r. Observe that mark x falls between marks qy and (q+1)y. Clearly remainder r is less than divisor y.

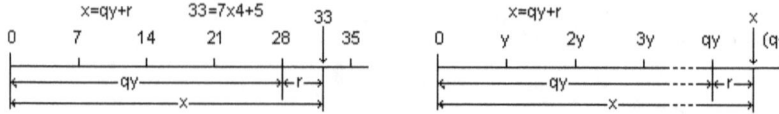

6 Fractions

Any number is represented by a point on the number line. *Fraction* is the name of a number that may not be an integer. Nevertheless, fraction is a number that is a point on the number line. The *fraction*, a *rational number*, is written as *integer m over integer n*.

$\dfrac{m}{n}$ *is a simple fraction where m and n are any integers and n ≠ 0.*

Important: in this chapter fraction components m and n are integers.

Fractions appeared when we studied division. However we did not call the remainder/divider expressions fractions.

A fraction can be located on the number line by using a straightforward geometrical construction (Section 6.1) that divides a line segment into n equal lengths. Integers mark off unit lengths on the number line. The geometrical construction shows how to divide any unit length into any number of equal lengths such as n=13, which shows there are 13 numbers in the length from marks 7 to 8 for example. In fact there is an infinity of numbers between any two integer marks (that's another subject). Fractions are some of that infinity of numbers, where we use infinity as an alias for *a number that is as large as we please.*

m may not be divisible by n   We say a number m is not divisible by a number n when the result cannot be written as an integer. Not divisible means dividing m by n produces a quotient plus remainder, which produces a fraction such as 13/23.

$$\frac{dividend}{divisor} = quotient + \frac{remainder}{divisor} \qquad \frac{4521}{23} = 196 + \frac{13}{23} \quad and \quad 4521 = 196\times23+13$$

The remainder is zero when m is divisible by n. For example

$$\frac{dividend}{divisor} = quotient + \frac{remainder}{divisor} \qquad \frac{4508}{23} = 196 + \frac{0}{23} \quad and \quad 4508 = 196\times23+0$$

fraction operations   Fractions are numbers. Therefore fractions can be manipulated by the operations addition, multiplication, subtraction, and division.

Arithmetic

6.1 Geometrical Construction of Equal Lengths

We show how to divide a unit line length into any number of equal lengths such as 5 equal fractional lengths. We employ the well known method of sliding a triangle. Actual unit length, 1 inch, 20 cm, etc., does not matter. What matters is that all fractional lengths are identical. You will need a triangle larger than shown. Try it and you will catch on immediately.

Draw horizontal line AB. Select arbitrary point mark 5 (see left side of the triangle in the Figure). This defines the unit length as the distance from 0 to mark 5. Draw line AC. On line AC use a compass and mark off points 1, 2, 3, 4, and 5 so that they are spaced equally. The length is arbitrary.

Place a 30/60/90 triangle and ruler as shown so that the side of the triangle is touching points AC5 and arbitrarily located point 'mark 5'. Use the left side of the triangle as a straight edge. Draw a line from AC point 5 through AB point 'mark 5', which becomes AB point 5.

Hold ruler down firmly, while sliding triangle up until triangle side touches AC point 4. Draw line from AC point 4 until it crosses line AB. The crossing point becomes AB point 4, and line AC4/AB4 is parallel to line AC5/AB5.

Repeat for points 3, 2, 1. Lengths between AB points 0, 1, 2, 3, 4, and 5 are fractional lengths 1/5, because they divide the length from A_0 to A_5 into five equal lengths.

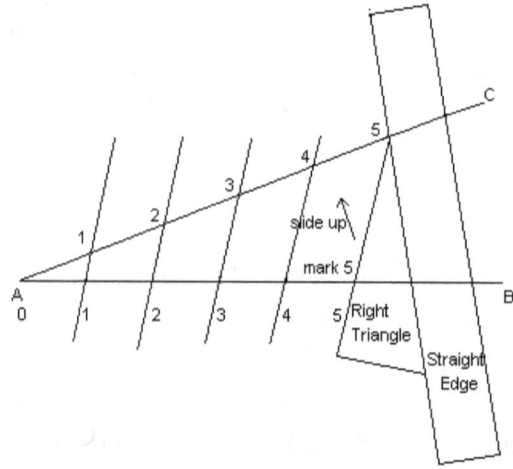

6.2 Fractions and the Number Line

Any number is a point on the number line. Integers are represented by marks on the number line, which are separated by equal unit lengths.

Use the geometrical construction (6.1) to divide each unit length into 7 equal lengths to produce a number line we label with two related scales.

The Standard Division Algorithm shows that 7 divided by 7 equals 1, and 14/7=2, 21/7=3. Now we can label the lower scale as shown, which we know is correct at integer marks 1, 2, 3. How about the other (fraction) marks?

The other marks are equidistant, so let the length equal x. Seven lengths x equal the length from 0 to 1, or 1 to 2, and so forth. The unit length equals the sum of seven 1/7 lengths (7 copies of length 1/7).

$$7x = 1 \quad \Rightarrow \quad divide\ both\ sides\ by\ 7 \rightarrow \frac{7x}{7} = \frac{1}{7} \quad \Rightarrow \quad x = \frac{1}{7}$$

$$mx = m\frac{1}{7} = \frac{1}{7} + \frac{1}{7} + \frac{1}{7} + \cdots + \frac{1}{7} = \frac{1}{7}(1+1+1+\cdots+1) = \frac{m}{7} \quad (m\ lengths\ \frac{1}{7})$$

The length 1/7 is the length from 0 to mark 1/7. Then 7 lengths 1/7 is the length from 0 to mark 7/7, which is the same as the length from 0 to 1. We can generalize from here and say *m lengths 1/n is the length from 0 to m/n*.

Fraction Definition Given positive integers m, n with n>0. Divide every unit line segment, such as from mark 56 to mark 57, into n equal lengths. This divides the number line from 0 to infinity into lengths 1/n. Then label the first mark 0 as 0/n, the next 1/n, 2/n, ..., m/n, etc. This set of labels representing lengths are referred to as fractions.
Fraction m/n is a quantity m of lengths 1/n (m/n = m × 1/n).

Proposition 1: *If a line of length m is divided into n equal parts, then the length of one part is* $\dfrac{m}{n}$

On the number line divide each unit into n equal parts 1/n, then

$$1 \text{ unit } length = \frac{n}{n} = \frac{1}{n} + \frac{1}{n} + \cdots + \frac{1}{n} = n \times \frac{1}{n} = a \text{ quantity } n \text{ of } 1/n \text{ parts}$$

$$m \times 1 = m \times n \times (1/n)$$

$$m = n \times m \times (1/n) = n \times (m/n)$$

$$m = a \text{ quantity } n \text{ of } m/n \text{ parts}$$

$$\text{so that the length of one part is } m/n$$

Proposition 2: *If* $\dfrac{mk}{nk} = \dfrac{m}{n}$, *then the two fractions are the same point on the number line.*

On the number line divide each unit into n equal parts, and
divide each length 1/n into k equal parts, then

$$m/n = m \times (1/n) \quad and \quad (1/n) = k \times (1/nk)$$

$$m/n = m \times k \times (1/nk)$$

$$m/n = mk \times (1/nk) = mk/nk$$

Names Names for a fraction's numbers are numerator and denominator.

$$fraction = \frac{numerator}{denominator} \quad such \ as \quad x = \frac{783}{147}$$

Where is the point on the number line? The usual number line marks are the integers. Where is the point representing the number 73/11? Well, we know from the division algorithm that mark 1 equals mark 11/11, mark 6 = 66/11, and mark 7=77/11. *This motivates us to interpret 73/11 as a division problem.*

$$\frac{73}{11} = 6 + \frac{7}{11} \quad by \ the \ division \ algorithm$$

The point on the number line for 73/11 is between marks 6 and 7, which we divide into 11 equal lengths. Now we can say the point on the line for 73/11 is seven 1/11 lengths to the right of mark 6.

6.3 Comparing Fractions

Two fractions are two points on the number line. The fraction to the left on the line is less than the fraction to the right. Equal fractions are fractions that represent the same point on the number line. Equal fractions are also referred to as equivalent fractions. We prefer equal. When we multiply m/n by 1 the fraction's value does not change. Then if we replace the 1 by the fraction k/k and multiply, a fraction km/kn equal to m/n is created (proposition 2).

$$\frac{m}{n} = \frac{1 \times m}{1 \times n} = \frac{k \times m}{k \times n} = \frac{km}{kn} \quad for\ example \quad \frac{7}{23} = \frac{1 \times 7}{1 \times 23} = \frac{6 \times 7}{6 \times 23} = \frac{42}{138}$$

Is 3/7 less than or greater than 5/11? The number line tells us its less.

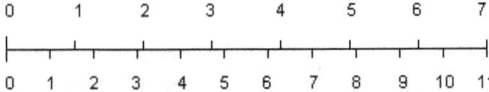

Is fraction m/n less than (<), equal to (=), or greater than (>) fraction p/q? Comparing 3/7 and 5/11 is comparing apples and oranges. We can only compare apple-oranges to apple-oranges (fractions with the same denominators). Convert apples and oranges to apple-oranges by multiplying numerators and denominators as shown .

$$\frac{3}{7} = \frac{3}{7} \times \frac{11}{11} = \frac{33}{77} \quad and \quad \frac{5}{11} = \frac{5}{11} \times \frac{7}{7} = \frac{35}{77} \quad and \quad 33 < 35 \quad \rightarrow \quad \frac{3}{7} < \frac{5}{11}$$

Replacing numbers by variables, we produce a general solution. Convert apples (1/n) and oranges (1/q) to apple-oranges (1/nq) or (1/qn). Then compare mq to pn.

$$\frac{m}{n} = \frac{m}{n} \times \frac{q}{q} = \frac{mq}{nq} \quad and \quad \frac{p}{q} = \frac{p}{q} \times \frac{n}{n} = \frac{pn}{qn}$$

The key to comparisons is to form equal denominators such as nq.

$$if\ the\ fractions\ are \frac{m}{n} = \frac{3}{7} and \frac{p}{q} = \frac{5}{11}, then\ form \frac{mq}{nq} = \frac{33}{77} and \frac{pn}{qn} = \frac{35}{77}$$

Now ask is mq < , = ,or > pn?   *Is 33 <, = ,or > 35?*

6.4 Fraction Addition

Integers may be added if they have the same units of length. We cannot add 3 yards to 2 feet, but we can add 9 ft (3 yards) to 2 ft. The same is true of fractions. *One over the denominator n (1/n) is the fraction unit of length.* This is also true for integers where each integer n equals fraction n/1. Denominator 1 is the integer unit of length.

Fraction m/n is a quantity m of lengths 1/n (m/n = m × 1/n). Fraction 3/5 is a quantity 3 of lengths 1/5 (3/5 = 3 × 1/5). Fraction 11/17 is a quantity 11 of lengths 1/17 (11/17 = 11 × 1/17). We can add 3/5 (apples) to 11/17 (oranges) if we convert both to quantities of apple-oranges 1/(5×17)=1/85.

> *The key to fraction calculations is "use the same units of length."*

Same denominators Fraction addition is straightforward when you want to add two fractions with the *same denominator d*, because each fraction is a quantity of the same length 1/d. In other words, calculations are facilitated, because the fraction 1/d with unit of length 1/d (an apple-orange) can be factored out. For example factor out 1/7:

$$\frac{4}{7}+\frac{2}{7}=4\times\frac{1}{7}+2\times\frac{1}{7}=(4+2)\times\frac{1}{7}=6\times\frac{1}{7}=\frac{6}{7}$$

If you interpret the fractions as 4 each 1/7 and 2 each 1/7, then you have a quantity of 6 each 1/7. Consequently you can use the shortcut: *if the denominators are equal, then add the numerators.* Verify on the number line by starting from 0 and laying out the lengths 4/7, and then 2/7.

We repeat. Observe that any fraction m/n = m × 1/n (e.g. 29/734 = 29 × 1/734). You can say the fraction m/n is a quantity m of the fraction 1/n. Thus you add quantities p and q of 1/n to get (p+q)/n. In general

$$\frac{p}{n}+\frac{q}{n}=p\frac{1}{n}+q\frac{1}{n}=(p+q)\frac{1}{n}=\frac{p+q}{n} \quad e.g. \quad \frac{29}{734}+\frac{214}{734}=\frac{29+214}{734}=\frac{243}{734}$$

Different denominators Fraction addition is more complex when adding two fractions with *different denominators*, because this means adding quantities of different length such as 1/n and 1/q. The key step is converting both fractions to fractions that have the same unit of length 1/denominator (proposition 2 page 70).

$$\frac{m}{n}+\frac{p}{q}=\frac{m}{n}\times\frac{q}{q}+\frac{p}{q}\times\frac{n}{n}=\frac{mq}{nq}+\frac{pn}{qn}=mq\frac{1}{nq}+pn\frac{1}{qn}=(mq+pn)\frac{1}{nq}=\frac{mq+pn}{nq}$$

For example create two equal fractions each with a denominator equal to 12.

$$\frac{2}{3}+\frac{1}{4}=\frac{2}{3}\cdot\frac{4}{4}+\frac{1}{4}\cdot\frac{3}{3}=\frac{8}{12}+\frac{3}{12}=\frac{1}{12}(8+3)=\frac{11}{12}$$

Another example

$$\frac{7}{53}+\frac{22}{49}=\frac{7\cdot49}{53\cdot49}+\frac{22\cdot53}{53\cdot49}=\frac{343}{2597}+\frac{1166}{2597}=\frac{343+1166}{2597}=\frac{1509}{2597}$$

In other words, *cross multiply*. Multiply numerator 3 times denominator 7, and add the product of numerator 2 times denominator 8. Then multiply denominator 7 times denominator 8.

$$\frac{3}{8}+\frac{2}{7}=\frac{3\times7+2\times8}{7\times8}=\frac{21+16}{7\times8}=\frac{37}{56}$$

Emphasis: Study the following m and n details carefully to understand why we can cross multiply and avoid the details. The general case is

$$use\ \frac{m}{m}=\frac{n}{n}=1\ to\ add\ fractions\ with\ different\ denominators$$

$$\frac{b}{n}+\frac{c}{m}=\frac{b}{n}\times1+\frac{c}{m}\times1=\frac{b}{n}\times\frac{m}{m}+\frac{c}{m}\times\frac{n}{n}=\frac{bm}{nm}+\frac{cn}{mn}=(bm+cn)\frac{1}{mn}=\frac{bm+cn}{mn}$$

Extra factors The following example brings up an important point. Eextra factors can be introduced unintentionally. How using the *gcd* finds extra factors is explained in Section 6.8 page 78.

$$\frac{5}{6}+\frac{4}{9}=\frac{5}{6}\cdot\frac{9}{9}+\frac{4}{9}\cdot\frac{6}{6}=\frac{45+24}{54}=\frac{69}{54}$$

$$however\ \frac{69}{54}=\frac{3\cdot23}{3\cdot18}=\frac{23}{18}$$

6.5 Fraction Multiplication

Integer times integer Multiplication replaces sums with products.
$$m \times s = s + s + s + \cdots + s \quad (m\,times)$$

Integer times fraction multiplication also replaces sums with products such as the sum of m lengths 1/n. Clearly multiplication by an integer means addition of many copies of a number such as a fraction.

$$m \times \frac{1}{n} = \frac{1}{n} + \frac{1}{n} + \frac{1}{n} + \cdots + \frac{1}{n} \ (m\,times) = \frac{1+1+1+\cdots+1}{n} = \frac{m}{n}$$

What can a fraction multiplied by a fraction mean? Consider the area of the unit square with sides of length 1. The area of the unit square is 1×1=1. Divide the vertical y axis into 2 equal parts, and divide the horizontal x axis into 5 equal parts. This divides the unit square into ten equal areas whose total area equals 1. Therefore each of the ten equal areas have area 1/10, because a quantity 10 of 1/10 equals 10/10=1 the total area.

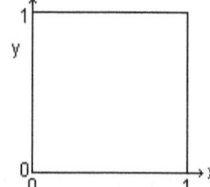 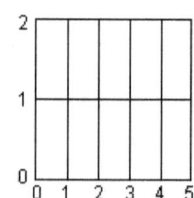 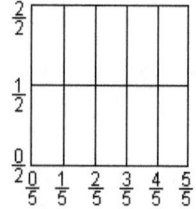

Each of the 10 areas has sides of length 1/2 and 1/5. The area of a rectangle is the product of its sides, and so each small area has area equal to 1/2 × 1/5. Therefore 1/2 × 1/5 must equal 1/10.

$$\frac{1}{2} \times \frac{1}{5} = \frac{1 \times 1}{2 \times 5} = \frac{1}{10}$$

What does 3/8 × 4/5 mean? For example the area of the 3/8 × 4/5 rectangle equals the area of 12 small areas where each small area has sides 1/8 and 1/5. Each small area equals 1/40, because

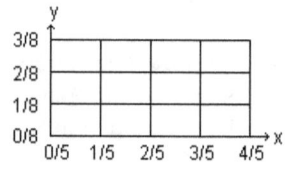

$$\frac{1}{8} \times \frac{1}{5} = \frac{5}{8 \cdot 5} \times \frac{8}{5 \cdot 8} = \frac{40}{40 \times 40} = \frac{1}{40} \implies area = 12 \times \frac{1}{40} = \frac{12}{40} = \frac{3 \times 4}{8 \times 5} = \frac{3}{8} \times \frac{4}{5}$$

Emphasis: Multiplying a fraction by 1 does not change its value. And a 1=n/n where n is any number. To understand this cancel the n's to be back to 1. For example

$$\frac{1}{5} = \frac{1 \times 7}{5 \times 7} \quad \text{because if the 7's are cancelled we are back to } \frac{1}{5}$$

In general

$$\frac{1}{n} = \frac{1 \times p}{n \times p} = \frac{p}{np} \quad \text{cancel the p's to return to } \frac{1}{n}$$

The examples show that a product of fractions is product of numerators over product of denominators.

$$\frac{m}{n} \times \frac{p}{q} = \frac{mp}{nq} = \frac{product\ of\ numerators}{product\ of\ denominators}$$

Here is how to multiply a fraction by an integer N.

$$N = \frac{N}{1} \quad \rightarrow \quad N \times \frac{p}{d} = \frac{N}{1} \times \frac{p}{d} = \frac{Np}{1d} = \frac{Np}{d}$$

A time saver cancels known common factors before multiplying. This saves a lot of work. For example:

$$\frac{12}{17} \times \frac{51}{28} = \frac{4 \cdot 3}{17} \times \frac{3 \cdot 17}{4 \cdot 7} = \frac{3 \cdot 3}{7} = \frac{9}{7}$$

A convention Fractions greater than 1 are referred to as *improper* fractions. There is a conventional writing of fractions greater than 1 that is not useful for calculations. The conventional writing is *not* a number based on position and weight (try adding two of them).

The convention is that you will see $2 + \frac{2}{7}$ *written as* $2\frac{2}{7}$

To get a number do this $2\frac{2}{7} = 2 + \frac{2}{7} = 2 \times \frac{7}{7} + \frac{2}{7} = \frac{14}{7} + \frac{2}{7} = \frac{16}{7}$

We prefer to use the original fraction 16/7. We avoid using the convention, which leads to errors too easily.

6.6 Fraction Subtraction

Fraction subtraction is based on the idea of same units of length, which is used in fraction addition. A minus sign replaces the plus sign. In effect that is the only difference. For example:

$$\frac{4}{7} - \frac{2}{7} = \frac{1}{7} \cdot \frac{4}{1} - \frac{1}{7} \cdot \frac{2}{1} = \frac{1}{7} 4 - \frac{1}{7} 2 = \frac{1}{7}(4-2) = \frac{1}{7} 2 = \frac{2}{7}$$

Again, you can factor out the denominator every time if you create equivalent fractions. For example:

$$\frac{2}{3} - \frac{1}{4} = \frac{2}{3} \cdot \frac{4}{4} - \frac{1}{4} \cdot \frac{3}{3} = \frac{8}{12} - \frac{3}{12} = \frac{1}{12}(8-3) = \frac{5}{12}$$

Another way to get the answer saves some writing.

$$\frac{2}{3} - \frac{1}{4} = \frac{2}{3} \cdot \frac{4}{4} - \frac{1}{4} \cdot \frac{3}{3} = \frac{8-3}{12} = \frac{5}{12}$$

Another example brings up the same important point.

$$\frac{5}{6} - \frac{4}{9} = \frac{5}{6} \cdot \frac{9}{9} - \frac{4}{9} \cdot \frac{6}{6} = \frac{45-24}{54} = \frac{21}{54}$$

$$\textit{however} \quad \frac{21}{54} = \frac{3 \cdot 7}{3 \cdot 18} = \frac{7}{18}$$

Sometimes one fraction is already in final form, and the result may be negative.

$$\frac{3}{16} - \frac{1}{4} = \frac{3}{16} - \frac{1}{4} \cdot \frac{4}{4} = \frac{3-4}{16} = -\frac{1}{16}$$

6.7 Fraction Division

Fraction division calculations are totally different from integer division calculations. We show that the quotient of two fractions (m/n)/(p/q) is the fraction equal to the product of m/n and the *reciprocal* q/p of p/q. Formally

$$\frac{\frac{m}{n}}{\frac{p}{q}} = \frac{\frac{m}{n}\times\frac{q}{p}}{\frac{p}{q}\times\frac{q}{p}} = \frac{\frac{m}{n}\times\frac{q}{p}}{1} = \frac{m}{n}\times\frac{q}{p} = \frac{mq}{np}$$

So that once you have some experience.

$$\frac{\frac{num_1}{denom_1}}{\frac{num_2}{denom_2}} = \frac{\frac{11}{17}}{\frac{4}{3}} = \frac{\frac{11}{17}\cdot\frac{4}{3}}{\frac{4}{3}\cdot\frac{3}{4}} = \frac{11}{17}\cdot\frac{3}{4} = \frac{33}{68} = \frac{num_1\cdot denom_2}{denom_1\cdot num_2}$$

A general argument for given two fractions x, y how do we calculate z=x/y?

$$let\ x = \frac{m}{n},\ \ y = \frac{p}{q},\ \ x = yz$$

$$the\ inverse\ of\ y\ is\ written\ as\ y^{-1}\ where\ y^{-1} = \frac{1}{y}$$

$$y^{-1}x = y^{-1}yz = z\ \ because\ \ y^{-1}y = \frac{1}{y}\times y = 1$$

$$z = xy^{-1} = \frac{m}{n}\times\frac{q}{p} = \frac{mq}{np}$$

Here are all of the steps, which you will actually skip once you have some experience.

$$\frac{\frac{a}{b}}{\frac{c}{d}} = \frac{\frac{a}{b}\cdot\frac{d}{c}}{\frac{c}{d}\cdot\frac{d}{c}} = \frac{\frac{a}{b}\cdot\frac{d}{c}}{1} = \frac{ad}{bc}$$

$$\frac{\frac{11}{17}}{\frac{4}{3}} = \frac{\frac{11}{17}\times\frac{4}{3}}{\frac{4}{3}} = \frac{\frac{11}{17}\times\frac{3}{4}}{\frac{4}{3}\times\frac{3}{4}} = \frac{\frac{11}{17}\times\frac{3}{4}}{1} = \frac{11}{17}\times\frac{3}{4} = \frac{33}{68}$$

Arithmetic

6.8 Greatest Common Divisor (gcd)

If you know the factors, a fraction can be simplified by *cancellation of the common factors* such as 3 and 5 in this example:

$$\frac{m}{n} = \frac{585}{165} = \frac{3\cdot3\cdot5\cdot13}{3\cdot5\cdot11} = \frac{3\cdot13}{11}\cdot\frac{3\cdot5}{3\cdot5} = \frac{3\cdot13}{11}\cdot\frac{3}{3}\cdot\frac{5}{5} = \frac{39}{11}$$

This can be difficult. For example what are the common factors of 69 and 1345, or of 4085376 and 297034? The gcd answers the question.

> The *greatest common divisor* (gcd) of two integers m and n is defined as the largest integer which divides both m and n. The gcd of m and n divides m and n.

Euclid's algorithm provides you with the means to find the gcd of any two integers m and n. The method is based on the fact that the last step of a sequence of division steps produces a zero remainder (that this always happens was proven by Euclid). You may remember how to find the gcd more readily if you note that the *next step is executing division of the reciprocal of the remainder term of the present step.* For example: 165/90 is the reciprocal of 90/165 (see below). We execute Euclid's algorithm on the number pair 585, 165. *The last divisor* 15 is the greatest common divisor (gcd)

$$\frac{m}{n} = q + \frac{r}{n} \quad \text{where q is the quotient and r is the remainder}$$

$$\frac{m}{n} = q_1 + \frac{r_1}{n} \qquad \frac{585}{165} = 3 + \frac{90}{165} \qquad 585 = 165\cdot3 + 90$$

$$\frac{n}{r_1} = q_2 + \frac{r_2}{r_1} \qquad \frac{165}{90} = 1 + \frac{75}{90} \qquad 165 = 90\cdot1 + 75$$

$$\frac{r_1}{r_2} = q_3 + \frac{r_3}{r_2} \qquad \frac{90}{75} = 1 + \frac{15}{75} \qquad 90 = 75\cdot1 + 15$$

$$\frac{r_2}{r_3} = q_4 + \frac{r_4}{r_3} \qquad \frac{75}{15} = 5 + \frac{0}{15} \qquad 75 = 15\cdot5 + 0$$

The gcd of 585 and 165 is 15. Divide both numbers by the gcd to form the reduced fraction.

$$\frac{585}{15} = 39 \quad and \quad \frac{165}{15} = 11 \quad \Rightarrow \quad \frac{585}{165} = \frac{15\times39}{15\times11} = \frac{39}{11}$$

6.9 Least Common Multiple (lcm)

The *least common multiple* of two integers m and n is the smallest integer which can be divided by both m and n (m and n divide the lcm).

$$m = 6, \ n = 9 \quad \Rightarrow \quad lcm = 18$$

because 18 *is the smallest integer divisible by both* 6 *and* 9

The lcm of two integers m and n is calculated as follows:

$$lcm = \frac{m \cdot n}{\text{gcd } of \ m \ and \ n}$$

if $m = 6$ *and* $n = 9$, *then by the gcd algorithm*

$$\frac{9}{6} = 1 + \frac{3}{6} \quad \Rightarrow \quad \frac{6}{3} = 2 + 0 \quad \textit{so that the gcd} = 3$$

$$and \ lcm = \frac{m \cdot n}{\text{gcd } of \ m \ and \ n} = \frac{6 \cdot 9}{3} = \frac{3 \times 2 \cdot 3 \times 3}{3} = 3 \times 2 \times 3 = 18$$

A practical method avoids calculating the gcd if you can factor m and n. In other words:

the factors of m *and* n *are* $6 = 3 \cdot 2$ *and* $9 = 3 \cdot 3$
the factors common to 6 *and* 9 *are* $2 \cdot 3 \cdot 3$
so that the lcm is $2 \cdot 3 \cdot 3 = 18$

The *important point* is that the step reducing the fraction is eliminated if the fractions have *denominators equal to the least common multiple*. Use n/n.

$$\frac{5}{6} + \frac{4}{9} = \frac{5}{6} \cdot \frac{3}{3} + \frac{4}{9} \cdot \frac{2}{2} = \frac{15 + 8}{18} = \frac{23}{18}$$

Sometimes one fraction is already in final form.

$$\frac{3}{16} + \frac{1}{4} = \frac{3}{16} + \frac{1}{4} \cdot \frac{4}{4} = \frac{3 + 4}{16} = \frac{7}{16}$$

Arithmetic

6.10 Review

Divide a line into equal lengths, which we say represent integers. This is the number line. Then a geometrical construction is used to divide each integer length into n lengths 1/n so that the length n×1/n =1. The distance from 0 to point m/n represents the *fraction* m/n.

> *The key to fraction calculations is "use the same units of length."*

A fraction is written as m over n. The fraction is a new type of number, the *rational number*. Fractions such as 7/23 are not the only fractions.

$$fraction = \frac{numerator}{denominator} \quad such \ as \quad \frac{e^{i\theta}}{147} \quad or \quad \frac{7+i3}{22} \quad or \quad \frac{\pi}{2} \quad or \quad \frac{\sqrt{3}}{\pi}$$

A fraction represents division. The fraction bar has the same meaning as the division sign. The fraction bar means divide.

$\dfrac{47}{121}$ (47 *divided by* 121), $\dfrac{3}{8}$ (3 *divided by* 8)

Equivalent fractions are fractions that have the same value. They are equal fractions. Multiply a fraction by 1 to create an equal fraction. This works when 1 takes the form n/n such as using 4/4 to multiply 5/8 to produce 20/32.

Fraction Operations add, subtract, multiply and divide fractions according to the rules of integer arithmetic.

Greatest Common Divisor (gcd) The greatest common divisor of two integers m and n is the largest integer which divides both m and n. Use Euclid's algorithm to calculate the gcd.

Least Common Multiple (lcm) The least common multiple of two integers m and n is the smallest integer which can be divided by both m and n. The lcm of two integers m and n is calculated as follows:

$$lcm = \frac{m \cdot n}{(gcd \ of \ m \ and \ n)}$$

The *gcd* divides both m and n. The *lcm* is divided by both m and n.

Problems 6

Instructional Objectives 6: Fractions
 1. Be able to explain when a fraction such as m/n is an integer.
 2. Be able to explain the difference between improper and proper fractions.
 3. Be able to explain when fractions are equivalent.
 4. Be able to explain how to create equal fractions.
 5. Be able to explain how convert an integer into a fraction.
 6. Be able to explain what fraction addition requires of the fractions.
 7. Be able to add and subtract any two fractions.
 8. Be able to multiply and divide any two fractions.
 9. Be able to define the gcd, and how to calculate it.
10. Be able to define the lcm, and how to calculate it.

The Basic Idea

An object n is divided into 23 parts.
1. What is one part called?
2. How is it written?

A number = 23/145.
3. Into how many parts is the original object divided?
4. How many parts of the original object does the fraction represent?

Equal Fractions

Find fractions that have a denominator of 128 equal to the following fractions.
5. 1/4   6. 3/8   7. 7/16   8. 5/32   9. 17/64

Arithmetic

Find fractions equal to the following pairs of fractions that have a minimum common denominator. Hint - find factors of denominators.

10. $\dfrac{1}{4}$ $\dfrac{3}{5}$ 11. $\dfrac{1}{14}$ $\dfrac{3}{21}$ 12. $\dfrac{7}{15}$ $\dfrac{5}{12}$ 13. $\dfrac{11}{16}$ $\dfrac{3}{15}$ 14. $\dfrac{1}{7}$ $\dfrac{3}{8}$

For each fraction find quotient plus remainder fraction.

15. 8/8	16. 9/8	17. 10/8	18. 11/8	19. 12/8	20. 13/8
21. 14/8	22. 15/8	23. 42/7	24. 22/7	25. 12/3	26. 14/3
27. 30/6	28. 33/6	29. 65/21	30. 27/11	31. 15/11	32. 19/4
33. 23/14	34. 45/9	35. 53/9	36. 27/5		

Change to larger denominators

37. 2/7 to 28ths	38. 3/5 to 20ths	39. 9/8 to 32ths	40. 3/10 to 1000ths
41. 11/6 to 72ths	42. 7/13 to 65ths	43. 2/9 to 27ths	44. /12 to 24ths

Find the missing numbers.

45. 46. 47. 48, 49. 50. 51.

$\dfrac{42}{7}=\dfrac{}{28}$ $\dfrac{2}{3}=\dfrac{}{27}$ $\dfrac{9}{4}=\dfrac{36}{}$ $\dfrac{}{6}=\dfrac{25}{30}$ $\dfrac{30}{}=\dfrac{5}{3}$ $\dfrac{33}{}=\dfrac{3}{7}$ $\dfrac{8}{21}=\dfrac{48}{}$

Fraction Addition

Add the following pairs of fractions. Hint - factor the denominators.

52. 53. 54. 55. 56. 57.

$\dfrac{8}{9}+\dfrac{1}{5}$ $\dfrac{3}{7}+\dfrac{1}{4}$ $\dfrac{5}{9}+\dfrac{3}{5}$ $\dfrac{9}{13}+\dfrac{2}{39}$ $\dfrac{2}{21}+\dfrac{3}{7}$ $\dfrac{4}{14}+\dfrac{1}{3}$

Convert to fractions and add.

58. 59. 60. 61. 62. 63.

$2\dfrac{7}{21}+4\dfrac{5}{14}$ $3\dfrac{1}{12}+5\dfrac{3}{18}$ $1\dfrac{3}{8}+3\dfrac{5}{6}$ $6\dfrac{7}{13}+1\dfrac{3}{26}$ $\dfrac{7}{16}+1\dfrac{11}{12}$ $4\dfrac{3}{11}+1\dfrac{2}{9}$

Greatest Common Divisor (gcd)

Find the gcd of each pair of numbers.

64.	65.	66.	67.
255 *and* 153	336 *and* 280	136 *and* 255	105 *and* 168

Least Common Multiple (lcm)

Find the lcm of each pair of numbers.

68.	69.	70.	71.
255 *and* 153	336 *and* 280	136 *and* 255	105 *and* 168

Add the following pairs of fractions using the lcm

72. $$\frac{13}{255}+\frac{7}{17}$$
73. $$\frac{4}{255}+\frac{5}{153}$$
74. $$\frac{9}{336}+\frac{2}{7}$$
75. $$\frac{13}{336}+\frac{3}{140}$$

Find the lcm of

76. $$\frac{1}{4}\ \frac{3}{5}\ \frac{7}{9}$$
77. $$\frac{2}{3}\ \frac{1}{15}\ \frac{9}{10}$$
78. $$\frac{5}{6}\ \frac{1}{2}\ \frac{1}{36}$$
79. $$\frac{3}{8}\ \frac{2}{9}\ \frac{1}{5}$$

Fraction Subtraction

Subtract the fractions using lcm

80. $$\frac{13}{85}-\frac{1}{17}$$
81. $$\frac{37}{51}-\frac{5}{255}$$
82. $$\frac{1}{14}-\frac{7}{168}$$
83. $$\frac{13}{56}-\frac{3}{140}$$

84. Decrease the value of 3/4 by twelve thirty ninths.
85. Decrease the value of 7/10 by two fifths.
86. Decrease the value of 4/9 by one fifth.
87. Decrease the value of 3/7 by one tenth.
88. Decrease the value of 17/32 by three eighths.
89. Decrease the value of 5/16 by two sevenths.

Arithmetic

Fraction Multiplication

Multiply the fractions. Reduce to lowest terms. Hint - factor all numbers.

90.	91.	92.	93.
$\dfrac{3}{22} \cdot \dfrac{55}{17}$	$\dfrac{18}{34} \cdot \dfrac{51}{28}$	$\dfrac{84}{13} \cdot \dfrac{39}{132}$	$\dfrac{28}{57} \cdot \dfrac{76}{7}$

94. Increase the value of 3/4 by four times.
95. Increase the value of 3/10 by two and one-half times.
96. Increase the value of 4/9 by five times.
97. Increase the value of 17/32 by two and two thirds times.
98. Increase the value of 4/9 by four and one half times.
99. Increase the value of 12/19 by one and one half times.

Find fraction to multiply by that decreases original fraction
100. Decrease 1/3 to 1/7.
101. Decrease 3/5 to 1/10.
102. Decrease 17/32 to 1/16.
103. Decrease 53/120 to 1/3.
104. Decrease 5/9 to 5/13.
105. Decrease 2/7 to 3/100.
106. Find 1/3 of 1/9
107. Find 2/7 of 9/11

Fraction Division
Divide the fractions. Reduce to lowest terms.

108	109	110	111
$\dfrac{\dfrac{3}{22}}{\dfrac{17}{55}}$	$\dfrac{\dfrac{18}{34}}{\dfrac{28}{51}}$	$\dfrac{\dfrac{13}{84}}{\dfrac{39}{132}}$	$\dfrac{\dfrac{57}{28}}{\dfrac{76}{7}}$

7 Decimals

Integers are whole numbers ranging in value from minus infinity to 0 to plus infinity, where infinity is an alias for a number as large as we please. Integers are represented by marks on the number line. Marks are separated by a distance equal to 1, as shown in this piece of the number line. The number line graphic shows that any number, such as x, is the sum of an integer (such as 4) and a part that is less than 1. The distance between the mark 4 and x is less than 1.

The integers mark off equal steps from minus infinity to positive infinity on the number line. The numbers in the gaps, such as x, between the integers 4 and 5 are *real* numbers, and so are the integers. The integers are a small subset of the real numbers. Some of the numbers in the gaps, such as 27+12/234, are fractions. Irrational numbers, such as √2, the square root of 2, fill in the remaining gaps. The numbers in the gaps have an integer part and a part that is less than 1. These numbers are written in different ways. We need a one way to write *any* number, which turns out to be the *decimal*. For example

$$27 + \frac{12}{234} = 27.05128.... \qquad \sqrt{2} = 1.41421....$$

A decimal representing any number has an integer part and a part that is less than 1. This is achieved by the decimal format *integer-dot-part<1* such as 627.23. The dot separates the parts. What does dot 23 (.23) mean?

The position weights of integer digits are 1, 10, 100, and so forth. The integer number 73841 has a digit in positions 4, 3, 2, 1, and 0. The position weights are 10000, 1000, 100, 10, 1. Position weight *decreases* by a factor of 10 as we move *right* one digit at a time.

Pretend a position exists to the right of units position 0. Then the weight of this position must also decrease by a factor of ten, because this is a move to the right. Consequently the weight of this position has to be one tenth of 1 or the fraction 1/10, which is less than 1. One more move to the right has to be 1/10 of 1/10 or 1/100 (the product of 1/10×1/10). For example, the fraction 23/100=20/100+3/100. This fraction reduces to 2/10+3/100, which we rewrite as (2×1/10)+(3×1/100) in anticipation of writing a decimal number.

The decimal number representing the fraction 23/100 is written as 0.23. The *dot* before the fractional part 23 differentiates the fraction 0.23 from the integer 23. This *dot* notation is referred to as the *decimal point*.

The mathematical community has agreed to write fraction 23/100 as 0.23. The decimal number 0.23 means zero integer part plus a 23/100 fraction part. Observe that the part less than 1, .23, is written as if it is an integer using only the ten symbols 0 to 9. However the dot before the 23 changes the meaning to (2×1/10)+(3×1/100)=23/100.

Decimal point   The decimal point in 0.23, marks the boundary between integer 0 and fraction 23/100. *A decimal point is simply a marker.*

Position number extended   The digits in the number 43210.12345 are equal to the *magnitude* of their position number. Positions to the right of the decimal point have negative numbers that start from −1. Why this is so is explained in upcoming paragraphs.

Again, the Decimal point The decimal point in 43210.12345, marks the boundary between integer 43210 and the decimal fraction 0.12345. *The decimal point is simply a marker.*

Numbers such as 65.23 are called decimals.

Decimals are numbers you can add, subtract, multiply, and divide. There are many applications of decimals in science and commerce.

Read the decimal fraction 12.43 as "twelve and forty three one hundreds" (1/100). Or read it as "12 point 43."

> *The essential idea is the dot marking the point separating the integer part that is greater than or equal to 0, from the part that is 0 or less than 1 as in 5.321, 5.0, 0.0, 0.321*

7.1 Powers of Ten and the Base 10 Number system

An *exponent x* is a symbol written above, and on the right of, another symbol known as the *base b* as in b^x. The expression b^x is referred to as a power; specifically the xth power of b. All arithmetic operations apply to exponents. *The exponent can be any type of number, which we restrict to integers* in this chapter. Furthermore let b=10 so that we can discuss powers of ten such as 10^5.

Many famous physical constants are usually expressed as a number in the range 1 to 9.999... times a power of 10. Observe the negative exponents.

velocity of light	$c = 2.997925 \times 10^8 \, meters/second$
Avogadro's number	$N_A = 6.0225 \times 10^{23} mole^{-1}$
charge of the electron	$e = 1.60210 \times 10^{-19} Coulomb$
Planck's constant	$h = 6.62517 \times 10^{-34} Joule \cdot second$
Boltzman's constant	$k = 1.3805 \times 10^{-23} Joule/degree$

Position weights are 1, 10, 100, 1000, etc., which are products of tens. This means the weights can be represented by a power of ten such as 10^p where the position number p is the exponent, and the base is 10.

We are interested in powers of ten, because ten is the base of our decimal number system. We have expanded integers as the sum of the value of their digits. For example the whole number $596 = 500 + 90 + 6$

We know we can do this, because when we learned to count we learned that for 596

> Digit 6 is in position zero, and that a digit equal to 6 in position 0 represents a quantity of six 1's.

> Digit 9 is in position one, and that a digit equal to 9 in position 1 represents a quantity of nine 10's.

> Digit 5 is in position two, and that a digit equal to 5 in position 2 represents a quantity of five 100's.

The idea of position increases the value of a digit *to the left* by a factor of ten. This is why the weight of any position is a power of ten.

Arithmetic

Since 10 with exponent 0 equals 1, $10^0=1$, the *position number* of the units digit was given the number zero.

Consider a conceptual explanation of a zero or negative exponent.

exponents add : $\quad 10^{2+3} = (10\cdot10)\times(10\cdot10\cdot10) = 10^2 \times 10^3 = 10^5$

exponents subtract : $\dfrac{10^5}{10^2} = \dfrac{10\cdot10\cdot10\cdot10\cdot10}{10\cdot10} = 10^5 \times 10^{-2} = 10^{5-2} = 10^3$

specific case : $\dfrac{10^0}{10^1} = 10^{0-1} = 10^{-1} = \dfrac{1}{10} \quad \Rightarrow \quad \dfrac{10^0}{10^1} = \dfrac{1}{10} \quad \Rightarrow \quad 10^0 = 1$

This is why exponents of position weights equal position number. For integers:

position#	position wt	exponent	
0	$1 = 10^0$	0	*Note : position number increases by one*
1	$10 = 10^1$	1	*for each move to the left.*
2	$100 = 10^2$	2	
3	$1000 = 10^3$	3	

Pretend you are in position 3. Position number decreases by one as you move *to the right* from position to position 3, 2, 1, 0. You reach the units digit in position 0. Then the digit *to the right* of the units digit must be in position 0–1 or –1 (minus one). For fractions we have

position#	position weight	exponent
–1	$\dfrac{1}{10} = \dfrac{1}{10^1} = 10^{-1}$	–1
–2	$\dfrac{1}{100} = \dfrac{1}{10^2} = 10^{-2}$	–2
–3	$\dfrac{1}{1000} = \dfrac{1}{10^3} = 10^{-3}$	–3

Now merge whole number positions with fraction number positions.

position#	3	2	1	0	–1	–2	–3
weight	1000	100	10	1	$\dfrac{1}{10}$	$\dfrac{1}{100}$	$\dfrac{1}{1000}$
	10^3	10^2	10^1	10^0	10^{-1}	10^{-2}	10^{-3}
exponent	3	2	1	0	–1	–2	–3

7.2 Decimal point and position weight

The logic of powers of ten places the integer part to the left of the decimal point, and the fraction part to the right of the decimal point. The decimal point separates integer digits from fractional digits. Integer digits have values based on positive powers of ten, whereas fractional digits have values based on negative powers of ten. For example:

$$63.71 = 6 \cdot 10^1 + 3 \cdot 10^0 + 7 \cdot 10^{-1} + 1 \cdot 10^{-2}$$

A more complex example is the number 210.1234. In this made up number the digits correspond to the position numbers, which in fact are actually $-1, -2, -3, -4$ to the right of the decimal point. The position number *sign* is positive for digits to the left of the decimal point, and negative for digits to the right of the decimal point. We expand 210.1234 as a sum of terms, and convert each term to a powers of ten format.

$$210.1234 = 200.0 + 10.0 + 0.0 + 0.1 + 0.02 + 0.003 + 0.0004$$

Check by adding the terms. Align the decimal points so that digits

with same weight are added

200.0

  10.0

    0.0

    0.1

    0.02

    0.003

    0.0004

Convert terms of 210.1234 to powers of ten.

$$210.1234 = 2 \cdot 100 + 1 \cdot 10 + 0 \cdot 1 + 1 \cdot \frac{1}{10} + 2 \cdot \frac{1}{100} + 3 \cdot \frac{1}{1000} + 4 \cdot \frac{1}{10000}$$

$$210.1234 = 2 \cdot 100 + 1 \cdot 10 + 0 \cdot 1 + 1 \cdot \frac{1}{10^1} + 2 \cdot \frac{1}{10^2} + 3 \cdot \frac{1}{10^3} + 4 \cdot \frac{1}{10^4}$$

$$210.1234 = 2 \cdot 10^2 + 1 \cdot 10^1 + 0 \cdot 10^0 + 1 \cdot 10^{-1} + 2 \cdot 10^{-2} + 3 \cdot 10^{-3} + 4 \cdot 10^{-4}$$

The unwritten decimal point falls between the 0 and -1 power terms (positions 0 and -1). The *decimal point is not present* in the expansion as a sum of powers of ten, because it only marks the boundary between the whole and fraction parts when the number is written as a decimal.

Arithmetic

7.3 Conversions to Decimals

The decimal forms of numbers are essentially mandatory for computational purposes. A decimal is any number that includes a decimal point. A decimal is usually the sum of an integer and a fraction: e.g. 46.9835. The fraction or integer may be zero as in 25 = 25.0, or 1/2 = 0.5. A decimal point is the dot marking the boundary separating the integer part from the fraction part.

Convert an Integer Any integer can be converted to decimal form by adding zero value in the form of a dot and zero(s). For example 43 = 43.0 = 43.000, etc.

Convert a fraction to a terminating decimal A way to convert some, *but not all*, fractions into a decimal is to make the denominator a power of ten. We multiply the fraction by n/n where the integer n times the denominator is some power of ten. You may want to refer to Section 7.6 to understand the following two examples, because there we *move* the decimal point as we change powers of ten.

Create an equivalent fraction with power of ten denominator

$$\frac{1}{25} = \frac{1}{25} \times \frac{4}{4} = \frac{4}{100} = 4 \cdot 10^{-2} = 0.4 \cdot 10^{-1} = 0.04 \cdot 10^{0} = 0.04$$

Convert 8 to a power of ten. Multiply by $1 = \frac{n}{n} = \frac{125}{125}$

$$\frac{7}{8} = \frac{7}{8} \times \frac{125}{125} = \frac{875}{1000} = 875 \cdot 10^{-3} = 0.875$$

Convert a fraction by division Any fraction is converted to a decimal by the Standard Division Algorithm. The new item is keeping track of the decimal point, which is a straightforward procedure.

Convert fraction 5/8 to 0.6250 The remainder is zero after the fourth division, which means the decimal terminates or repeats the 0 in this example. The ellipses ... mean that the number 0, in this example, repeats forever.

Convert fraction 2/11 to 0.1818 The remainders alternate: 2, 9, 2, 9, etc. The non zero remainders means the division can continue forever. The quotient digits 1, 8 repeat forever.

***Convert fraction 1/7 to* 0.142857**   The remainders recycle through 1, 3, 2, 6, 4, 5. The non zero remainders means the division continues forever. The number 142857 repeats forever.

```
                              0.142857...
                           7)1.000000...
0.62500...    0.1818...     -0
8)5.00000...  11)2.0000...  1 0
-0            -0            -7
 5 0           2 0           30
-4 8          -11           -28
  20            90            20
 -16           -88           -14
   40            20            60
  -40            11           -56
    00            90            40
   -00           -88           -35
      0...          2...         50
                                -49
                                  1...
```

0.6250 is a terminating decimal
0.181818... is a repeating decimal.
0.142857142857... is a repeating decimal.

Note: *The decimal form of irrational numbers does not terminate, nor repeat.* Here are 9 digit approximations of three irrational numbers.

$\pi \approx 3.14159265$ $\sqrt{2} \approx 1.41421356$ $\sqrt{3} \approx 1.73205081$

Theory of the Repeating Decimal

Felix Klein[2] offers this argument.

Theorem of Fermat for every prime number p, except 2 and 5,

$$10^{p-1} \equiv 1 (\mathrm{mod}\ p)$$

this means $\dfrac{10^{p-1}-1}{p} = \dfrac{10^{p-1}}{p} - \dfrac{1}{p} = N$ *integer*

Another theorem from the theory of numbers is the theorem that the smallest exponent δ is either p–1 or a divisor of p–1.

$$\frac{10^{\delta}-1}{p} = N \quad \Rightarrow \quad \frac{10^{\delta}}{p} = N + \frac{1}{p} \quad \Rightarrow \quad \frac{10^{\delta}}{p} \ and \ \frac{1}{p} \ differ \ by \ an \ integer \ N$$

this means the fractional part of $\dfrac{10^{\delta}}{p}$ *equals* $\dfrac{1}{p}$

This is the key idea: *Decimal numbers that differ by an integer have equal decimal fraction parts.*

Think of 1/p as a decimal. Then $10^{\delta} \times$ 1/p only moves the decimal point δ positions to the right. Consequently the digits are not changed.

We assume the decimal fraction has a repeated *period*, or block, of δ digits.

Examples showing δ may be less than p as well as equal to p–1 (Section 7.4).

$$\frac{1}{3} = 0.33333.... \quad (\delta = 1)$$

$$\frac{1}{11} = 0.090909.... \quad (\delta = 2)$$

$$\frac{1}{7} = 0.142857142857.... \quad (\delta = 6)$$

---

[2] Felix Klein 1908, "Arithmetic, Algebra, Analysis" , ISBN 048643480X

7.4 Decimals to Fractions

The Standard Division Algorithm was used to show that any rational number, such as a fraction m/n, has one block of digits or a repeating block of digits in its decimal fraction.

Now we want to show that any decimal with a repeating block equals a rational number. The key idea stems from the fact that decimal numbers that differ by an integer have equal decimal fraction parts (such as 23.987 and 654.987).

Given a decimal fraction with a repeating block of δ digits multiply it by 10^δ to leave the fraction part unchanged thereby creating a decimal number that differs from the decimal fraction by an integer. Then subtract the original decimal fraction to get an integer (Theory of the Repeating Decimal). Three examples.

let $x = 0.33333....$ $(\delta = 1)$

$10^1 x = 3.33333....$ *integer 3 plus a fraction part that is not changed*

$10^1 x - x = 3.33333.... - 0.33333....$

$9x = 3.33333.... - 0.33333....$

$9x = 3$ $\Rightarrow$ $x = \dfrac{1}{3}$

let $x = 0.090909....$ $(\delta = 2)$

$10^2 x = 09.090909....$

$10^2 x - x = 9.090909.... - 0.090909....$

$99x = 9.090909.... - 0.090909....$

$99x = 9$ $\Rightarrow$ $x = \dfrac{1}{11}$

let $x = 0.142857....$ $(\delta = 6)$

$10^6 x = 142857.142857....$

$10^6 x - x = 142857.142857.... - 0.142857....$

$999999x = 142857.142857.... - 0.142857....$

$999999x = 142857$ $\Rightarrow$ $x = \dfrac{1}{7}$

Arithmetic

7.5 Decimal Operations

Decimals are numbers that can be operated on by the addition, subtraction, multiplication and division operators. When we added or subtracted integers we first aligned their right-hand edges.

4567   *right-hand edges aligned*
+234
———
4801

In fact we were aligning integers at their unwritten decimal points when we added or subtracted.

4567.0   *decimal points aligned*
+234.0
———
4801.0

The alignment by right-hand edge or (unwritten) decimal point is alignment by position, which is necessary in order to add digits with the same position weight. The alignment by decimal points is equivalent to alignment by position. This is why the introduction of decimal points does not affect calculations using integers. The introduction of decimal points extends calculations with integers to calculations with any real numbers.

Add and subtract After aligning decimal points, ignore the decimal points when adding or subtracting.

234.567   234.567
+98.7   −98.700
———   ———
333.267   135.867

Multiply Pretend the decimal points are not there and just multiply the two numbers. Then insert a decimal point in the answer s positions to the left. You determine s by counting digits to the right of the decimal points in both numbers. We show that this is correct by example.

Multiply 435.56 by 34.567 while ignoring the decimal points to get 1,505,600,252. There are 2 + 3 or 5 digits to the right of the decimal points, which makes s = 5. The answer is 15056.00252. Verify by rounding down and up to get an estimate of 435 × 35 = 15225, which is the same order of magnitude as 15056.

$$
\begin{array}{r}
43556 \\
\times 34567 \\
\hline
304892 \\
2613360 \\
21778000 \\
174224000 \\
+1306680000 \\
\hline
1505600252
\end{array}
$$

$$435.56 \times 34.567 = 15056.00252$$

Divide Convert the divisor to an integer and use conventional division. In the case of 234.567/98.7 convert 98.7 to 987. In other words multiply both numbers by ten, This moves the decimal point one place to the right in both numbers (see 7.6). Then divide, while pretending the decimal point is not there.

$$
\begin{array}{r}
2.37....... \\
987\overline{)2345.670000} \\
1974 \\
\hline
3716 \\
2961 \\
\hline
7557 \\
6909 \\
\hline
6480 \ etc
\end{array}
$$

7.6 Moving the decimal point left and right

***To the right means times* 10** If we rewrite the number 210.123 by moving the decimal point one place to the right, then the number changes to 2101.23. The rewritten value is ten times the original value.

$$2101.23 = 2 \cdot 10^3 + 1 \cdot 10^2 + 0 \cdot 10^1 + 1 \cdot 10^0 + 2 \cdot 10^{-1} + 3 \cdot 10^{-2}$$
$$= 10 \cdot (2 \cdot 10^2 + 1 \cdot 10^1 + 0 \cdot 10^0 + 1 \cdot 10^{-1} + 2 \cdot 10^{-2} + 3 \cdot 10^{-3})$$
$$= 10 \times 210.123$$

Moving the decimal point in any number, such as 210.123, one position to the right is the same as multiplying by ten.

For example:

Every multiplication by ten moves the decimal point to the right
$$5.0 \times 1000 = 50.0 \times 100 = 500.0 \times 10 = 5000.0 \times 1 = 5000$$

***To the left means times* 1/10** If we rewrite the number 210.123 by moving the decimal point one place to the left then the number changes to 21.0123. The rewritten value is one–tenth the original value.

$$21.0123 = 2 \cdot 10^1 + 1 \cdot 10^0 + 0 \cdot 10^{-1} + 1 \cdot 10^{-2} + 2 \cdot 10^{-3} + 3 \cdot 10^{-4}$$
$$= \frac{1}{10}(2 \cdot 10^2 + 1 \cdot 10^1 + 0 \cdot 10^0 + 1 \cdot 10^{-1} + 2 \cdot 10^{-2} + 3 \cdot 10^{-3})$$
$$= \frac{1}{10} \times 210.123$$

Moving the decimal point one position to the left in any number, such as 210.123, is the same as dividing by ten.

For example:

Every division by ten moves the decimal point to the left
$$\frac{500}{1000} = 500.0 \times \frac{1}{1000} = 50.0 \times \frac{1}{100} = 5.0 \times \frac{1}{10} = 0.5$$

7.7 Percent

Percent means per-hundred. The symbol for percent is %. *Percent is a decimal fraction equal to a fraction whose denominator is 100.* Therefore conversion of a fraction to percent is done by finding an equal fraction with a denominator equal to 100. Multiply by n/n=100/100 to achieve this result.

$$\frac{3}{8} = \frac{3}{8} \cdot \frac{100}{100} = \frac{300}{8} \cdot \frac{1}{100} = \frac{37.5}{1} \cdot \frac{1}{100} = \frac{37.5}{100} = 37.5\%$$

Another example with conversion back to decimal.

$$0.436 = \frac{0.436}{1} \cdot \frac{100}{100} = \frac{43.6}{100} = 43.6\%$$

$$percent \ \ to \ \ decimal : 43.6\% = \frac{43.6}{100} = 0.436$$

Yet another way to make the conversion is to divide first.

$$\frac{7}{31} = 0.225806.... = 0.225806... \times \frac{100}{100} = \frac{22.5806....}{100} \approx 22.58\%$$

Percent of any number Calculating the percent of any number is a multiplication problem. A simplifying step is to convert the percentage to an integer times a power of ten before multiplying. Finally, you convert to decimal format. For example:

$$43\% \ of \ 122.58 = 122.58 \times \frac{43}{100} = 12258 \times 10^{-2} \times 43 \times 10^{-2} = 12258 \times 43 \times 10^{-4}$$

$$
\begin{array}{r}
12258 \\
\times \quad 43 \\
\hline
36774 \\
490320 \\
\hline
527094 \\
\end{array}
$$

and $527094 \times 10^{-4} = 52.7094$

Check: 50% of 122 is 61. 40% of 122 is 48.8. Then 52.7094 makes sense.

7.8 Interest

This is a very important topic. Knowing how to compute the *actual or effective* interest rate in an offer or transaction can save you real money.

Read very carefully before you sign *any* document. Do not pay any attention to what is *said*. If you do it will probably cost you real money. The talk is usually intended to distract you. Talk is very low cost to the seller, and can be high cost to the buyer.

| Only the *written* contract terms matter. |

Earn the seller's respect by telling him that. If he does not stop talking leave! Seller's are like street cars. There is always another coming along.

Every tenth of a point is significant!

interest (**i**)   The amount paid for use of borrowed money.

principal (**p**)   The amount borrowed (loaned) or invested.

rate (**r**)   The percent of the principal paid as interest per specified time unit, usually one year.

time (**t**)   The time period money is borrowed or invested.

amount (**s**)   The principal plus interest earned.

Simple Interest Interest is earned only on the principal, and not on interest accrued.

$i = prt$ Example: simple interest earned on $500 principal, at a 7% rate per year, and for 18 months is
$i = p \times r \times t = 500$ dollars $\times$ 0.07/year $\times$ 1.5 years = $52.50.

$s = p + prt$ For the same example the amount due at the end of 18 months is $s = \$500 + \$52.50 = \$552.50$

Rules for partial years The period interest is earned excludes the first day, and includes the last day. The amount is due and payable on the last day.

Three methods for calculating interest The methods differ in the time period used.

1. *Time period-30 day months, Ordinary method* Assume for simplicity of calculation there are twelve months of 30 days each, or 360 days per year.

2. *Time period-actual days, Exact or accurate method* Count the actual number of days in the time period.

3. *Time period-months plus days, Banker's method* Time is expressed in months plus days or exact days. For example the time from May 7 to September 25 is 4 months (May 7 to September 7) plus 18 days (September 8 to 25).

Examples
You borrow $750 for 120 days at 9% per 365 day year.
Exact simple interest is $750 \times .09 \times 120/365 = \$22.19$.

Your credit card debt is $1000. Interest is 19% per year. Banker's simple interest accumulated *each month* is $\$1000 \times 0.19 \times 1/12 = \$15.83$.

Your auto loan is $15,000. Interest is 14% per year. Annual simple interest is $\$15,000 \times 0.14 = \$2100$.

Compound Interest Compound interest differs from simple interest, because *the principal is increased by the interest earned* during each interest period for the duration of the loan.

Compound amount s = original principal + compound interest

Frequency of conversion c is the number of times per year interest is converted into principal. The number of times is 365 (daily), 52 (weekly), or as agreed upon. A quarterly (c = 4) conversion period is typical.

Nominal rate of interest is the interest rate per year.

Effective rate of interest is the actual interest rate earned (paid) per year. Compounding increases the nominal rate to the effective rate.

Arithmetic

Example What is the effective annual interest rate of 9% compounded quarterly? The quarterly rate is 9/4 %. You will need a calculator here.

$$1+i_{eff} = \left(1+\frac{.09}{4}\right)^{4\times1} = 1.093083.... \Rightarrow i_{eff} = 9.3083....\%$$

Equation for time period t $\quad s = \left(1+\frac{r}{c}\right)^{c\times t}$

Example What amount is due at the end of 3 years on original principal of $5000 at a compound interest rate of 8% per year.

The amount at the end of any one year is $s = p + pr = p(1+r)$
$s_1 = 5000(1+.08) = 5400$
$s_2 = 5400(1+.08) = 5832$
$s_3 = 5832(1+.08) = 6298.56$ or
$s_3 = p(1+r)^t = 5000(1+.08)^3 = 5000\times1.259712 = 6298.56$

The power of compound interest *This is why you are advised to start saving/investing when you are young.* The ratio s/p is the growth over the years for one initial investment p.

t years	$\frac{s}{p} = (1.06)^t$	$\frac{s}{p} = (1.08)^t$	$\frac{s}{p} = (1.10)^t$
5	1.3382	1.4693	1.6105
10	1.7908	2.1589	2.5937
15	2.3966	3.1722	4.1772
20	3.2071	4.6610	6.7275
25	4.2919	6.8485	10.8347

7.9 Continued Fractions

A *continued fraction* (CF) is a fraction
  whose denominator contains a fraction,
  whose denominator contains a fraction, etc.

A CF has many applications in higher mathematics such as *approximation* of fractions, and irrational numbers such as π. A CF produces a sequence of approximations that converge very rapidly (we hope) to the exact value. For example, expand 9/38 into a CF with all numerators equal to one.

$$\frac{9}{38} = \frac{1}{\frac{38}{9}} = \frac{1}{4+\frac{2}{9}} = \frac{1}{4+\frac{1}{\frac{9}{2}}} = \frac{1}{4+\frac{1}{4+\frac{1}{2}}} \approx 0.2368421....$$

approximations $\quad \dfrac{9}{38} \approx \dfrac{1}{4+0} = 0.250 \qquad \dfrac{9}{38} \approx \dfrac{1}{4+\dfrac{1}{4+0}} = \dfrac{4}{17} = 0.23529$

Expand π into a CF to find approximations such as 355/113.

$$\pi = 3.141592653.... = 3+0.1415.... = 3+\frac{1}{7+0.0625....} = 3+\frac{1}{7+\frac{1}{15+0.99....}}$$

$$= 3+\cfrac{1}{7+\cfrac{1}{15+\cfrac{1}{1+0.0034....}}} = 3+\cfrac{1}{7+\cfrac{1}{15+\cfrac{1}{1+\cfrac{1}{293+0.096....}}}}$$

$$\pi \approx 3+\frac{1}{7} = \frac{22}{7} = 3.142857143 \qquad\qquad \pi \approx 3+\cfrac{1}{7+\cfrac{1}{15}} = \frac{333}{106} = 3.141509434$$

$$\pi \approx 3+\cfrac{1}{7+\cfrac{1}{15+\cfrac{1}{1}}} = \frac{355}{113} = 3.1415920$$

$$\pi \approx 3+\cfrac{1}{7+\cfrac{1}{15+\cfrac{1}{1+\cfrac{1}{293}}}} = \frac{104348}{33215} = 3.141592654$$

Arithmetic

The CF for the square root of 2 is

$$\sqrt{2} = 1 + \cfrac{1}{2 + \cfrac{1}{2 + \cfrac{1}{2 + \cfrac{1}{2 + \cfrac{1}{2 + ...}}}}}$$

This converges too slowly to be useful.

$$\sqrt{2} \approx 1 + \frac{1}{2} = \frac{3}{2} = 1.50 \qquad \sqrt{2} \approx 1 + \cfrac{1}{2 + \cfrac{1}{2 + \cfrac{1}{2 + \cfrac{1}{2}}}} = \frac{41}{29} = 1.41379$$

Approximate a fraction with large numbers.

$$\frac{547}{97348} = \cfrac{1}{177 + \cfrac{1}{1 + \cfrac{1}{29 + \cfrac{1}{2 + \cfrac{1}{1 + \cfrac{1}{1 + \cfrac{1}{3}}}}}}}$$

Work backwards to check the result

$$1 + \frac{1}{3} = \frac{4}{3} \quad \rightarrow \quad 1 + \frac{3}{4} = \frac{7}{4} \quad \rightarrow \quad 2 + \frac{4}{7} = \frac{18}{7} \quad \rightarrow$$

$$29 + \frac{7}{18} = \frac{529}{18} \quad \rightarrow \quad 1 + \frac{18}{529} = \frac{547}{529} \quad \rightarrow \quad 177 + \frac{529}{547} = \frac{97348}{547} \, qed$$

Approximations

$$\frac{547}{97348} = 0.005619016$$

$$\frac{1}{177 + 0} = 0.005649718$$

$$\cfrac{1}{177 + \cfrac{1}{1 + \cfrac{1}{29 + \cfrac{1}{2 + 0}}}} = \frac{61}{10856} = 0.005619013$$

7.10 Review

Decimal fraction Decimal fraction examples are 3.141592 and 23/1000. Some decimals are written as rational numbers with power of ten denominator (23/1000) or with a decimal point (0.023).

Decimal point The decimal point in 43210.1234, marks the boundary between integer 43210 and decimal fraction 0.1234. *The decimal point is simply a marker.*

The essential idea is the dot marking the point separating the integer part from the fraction part that is less than 1. This fractional part is not necessarily equivalent to some fraction that is the ratio of two integers.

Conversions to decimal Any integer can be converted to decimal form by adding dot zero(s). For example 43 = 43.0 = 43.000.

Division is the key to converting any fraction to decimal format. Any rational number *m/n* can be converted to decimal form by division.

Decimal point and position weight The decimal point separates integer digits from fractional digits. The integer part is to the left of the decimal point, and the fraction part is to the right of the decimal point. Integer digits have weights based on positive powers of ten, whereas fractional digits have weights based on negative powers of ten. For example:

$$63.71 = 6 \cdot 10^1 + 3 \cdot 10^0 + 7 \cdot 10^{-1} + 1 \cdot 10^{-2}$$

Decimal operations Decimals can be operated on by the addition, subtraction, multiplication, and division operators. When we added or subtracted integers we first aligned their right-hand edges. In fact we were aligning integers at their unwritten decimal points when we added, subtracted, etc.

4567	*right hand edges aligned*		4567.0	*decimal points aligned*
+234			+234.0	
4801			4801.0	

Arithmetic

To add or subtract align the numbers by their decimal points and proceed as if you are adding integers.

To multiply pretend the decimal points are not there, and just multiply the two numbers. Then insert a decimal point in the answer s positions to the left. Determine s by counting digits to the right of the decimal points in both numbers.

To divide make the divisor an integer and use conventional division. In the case of 234.567 divided by 98.7 make 98.7 into the integer 987. So move the decimal point in both numbers one place to the right. Then divide 2345.67/987 to get the quotient 2.3765.....

To add an integer to a decimal fraction align decimal points and add.

three and seven tenths $= 3 + 0.7 = 3.7$

twenty four and five thousandths $= 24 + 0.005 = 24.005$

two hundred ten and thirteen hundredths $= 210 + 0.13 = 210.13$

Moving decimal point left and right Multiplying by ten moves the decimal point one position to the right. Dividing by ten moves the decimal point one position to the left.

Percent Percent means per-hundred. The symbol for percent is %. Percent is a decimal fraction equal to a fraction whose denominator is 100. Therefore conversion of a fraction to percent is done by finding an equivalent fraction with a denominator equal to 100.

Interest Interest is rent for money loaned or borrowed. It has tremendous personal and commercial importance.

Problems 7

Instructional Objectives: Decimals
1. Be able to define a decimal fraction.
2. Be able to define a mixed number.
3. Be able to explain how to convert a fraction to a decimal fraction.
4. Be able to explain how to convert a mixed number to an integer plus a decimal fraction.
5. Be able to explain how to convert a fraction to a decimal fraction with a power of ten denominator.
6. Be able to explain what boundary the decimal point marks in a number.
7. Be able to explain how moving the decimal point left or right affects the value of a number.
8. Be able to explain how leading and trailing zeros affect the value of an integer.
9. Be able to explain how leading and trailing zeros affect the value of a decimal fraction.
10. Be able to show the power of ten representing tenths, hundredths, and thousandths.
11. Be able to define a rational number as a decimal fraction. Give examples.
12. Be able to define an irrational number as a decimal fraction. Give examples.
13. Be able to define percent as a decimal fraction.
14. Be able to calculate 27% of 342.
15. Be able to add and subtract mixed numbers in decimal format.
16. Be able to multiply mixed numbers in decimal format.
17. Be able to divide mixed numbers in decimal format.

Definitions
1. The denominator of a decimal fraction is a power of what number?
2. What is the purpose of the decimal point?
3. A mixed number is the sum of two parts. What are their names?
4. The decimal point separates two parts of any number. What are their names?

Arithmetic

Conversions to decimal form

Convert to a decimal.

5.	6.	7.	8.	9.	10.	11.	12.	13.	14.	15.
$\dfrac{3}{6}$	$\dfrac{3}{7}$	$\dfrac{5}{20}$	$\dfrac{13}{17}$	$\dfrac{17}{123}$	$\dfrac{5}{16}$	$\dfrac{1}{6}$	$\dfrac{1}{11}$	$\dfrac{13}{45}$	$\dfrac{23}{54}$	$\dfrac{45}{111}$

Convert the following phrases into decimals.
16. One-hundred-three one-thousandths
17. Thirty-nine one-thousandths
18. Nine one-hundredths

Decimal point and position weight

Expand the numbers into a sum of powers of ten. Omit the zero terms.

19.	20.	21.	22.
5.3	347.59	502.01	20001.0001

Moving decimal point left and right

23. What operation with what number moves a decimal point to the left?
24. What operation with what number moves a decimal point to the right?

Leading and trailing zeros

25. Adding a leading zero to the integer part with value x changes the value to what?
26. Adding a trailing zero to the integer part with value x changes the value to what?
27. Adding a leading zero to the fractional part with value x changes the value to what?
28. Adding a trailing zero to the fractional part with value x changes the value to what?

Tenths, hundreds, thousandths
29. Write thirty seven hundredths in numerical format.
30. Write thirteen tenths in numerical format.
31. Write fifty seven thousandths in numerical format.
32. Write one hundred ninety nine ten thousandths in numerical format.
33. Write one one millionth in numerical format.

Rational numbers and repeating decimals

34. Find the repeating decimal produced by 1/6.
35. Find the repeating decimal produced by 1/11.
36. Find the repeating decimal produced by 2/13.
37. Find the repeating decimal produced by 77/123.
38. Find the repeating decimal produced by 2/11.
39. Find the repeating decimal produced by 12/123.

Irrational numbers and non-repeating decimals
Use a calculator.
40. Find the non-repeating decimal produced by square root of 3.
41. Find the non-repeating decimal produced by $\pi/2$.
42. Find the non-repeating decimal produced by log 2
43. Find the non-repeating decimal produced by ln 2
44. Find the non-repeating decimal produced by e^2
45. Explain why the ratio of two integers cannot be irrational.

Percent

Convert to percent to 3 decimal places.

46.	47.	48.	49.	50.	51.
$\dfrac{1}{4}$	$\dfrac{3}{25}$	$\dfrac{33}{100}$	$\dfrac{12}{20}$	$\dfrac{1}{14}$	$\dfrac{3}{9}$

Percent of any number

Find the specified percent of numbers to 3 decimal places.

52.	53.	54.
22% of 123	7% of 9980	39% of 243

Interest

Find the missing number. simple interest $i = $p × r %/year × t years.
55. I = $?,   p = $500,   r = 9%,   t = 2 years
56. I = $200,   p = $?,   r = 7%,   t = 3 years
57. I = $1000, p = $3000, r = ?%,   t = 5 years
58. I = $700,   p = $2500, r = 7%,   t = ? years

Arithmetic

Find compound interest earned or paid.
59. I = $?, p = $1000, r = 7%,   t = 2 years, c = 1
60. I = $?, p = $1000, r = 7%,   t = 3 years, c = 1
61. I = $?, p = $1000, r = 7%,   t = 5 years, c = 4

Credit card (disaster)
62. I = $?, p = $1000, r = 18%, t = 1 years, c = 12

Investments
63. s = $?, p = $10000, r = 5%,   t = 20 years, c = 1
64. s = $?, p = $10000, r = 7%,   t = 20 years, c = 1

Decimal addition and subtraction

Find the sums and differences.

65.	66.	67.	68.
23.005 + 322.1	23.005 − 322.1	3007.1 + 1.3002	3007.1 − 1.3002
69.	70.	71.	72.
100.01 + 1.01	100.01 − 1.01	5.06 + 3.067	5.06 − 3.067
73.	74.		
90034.002 + 100.6	90034.002 − 100.6		

Decimal multiplication
Find the products.

75.	76.	77.	78.	79.
23.005 × 322.1	3007.1 × 1.3002	100.01 × 1.01	5.06 × 3.067	90034.002 × 100.6

Decimal division
Find the quotients

80 23.005 ÷ 322.1	81 3007.1 ÷ 1.3002	82 100.01 ÷ 1.01

83 5.06 ÷ 3.067   84 90034.002 ÷ 100.6

8 The Basic Laws of Operations

Mathematics is about the reasoning, the ideas, not just the procedures.

Felix Klein[1] enumerates the eleven laws all elementary reckoning can be based on. There are *five fundamental laws* upon which *addition* depends. For any numbers x, y, z

(1) $x + y$ *is always a number* (*addition is always possible*)

(2) $x + y$ *is one valued*

(3)   *The associative law holds* $(x + y) + z = x + (y + z)$

(4)   *The commutative law holds* $x + y = y + x$

(5)   *The monotonic law holds*   *If* $y > z$, *then* $x + y > x + z$

There are *five exactly analogous laws* upon which *multiplication* depends.

(6) $x \times y$ *is always a number* (*multiplication is always possible*)

(7) $x \times y$ *is one valued*

(8)   *The associative law holds* $(x \times y) \times z = x \times (y \times z)$

(9)   *The commutative law holds* $x \times y = y \times x$

(10) *The monotonic law holds*   *If* $y > z$, *then* $x \times y > x \times z$

Multiplication is connected to addition by the distributive law.
(11) $x \times (y + z) = x \times y + x \times z$

Klein claims "that it is easy to show that all elementary reckoning can be based on these eleven laws." We take his word for it, because proving this statement requires a very big digression. Klein does refer to the content of original sources, which one can pursue. There are many reasons for knowing and understanding the eleven laws. Here are two.

1. We use them all of the time. Most of the time we use them without our being aware we are doing that, because their application is not explicit.

2. These laws play a central role in arithmetic. They are hidden behind the scene, so to speak, of all operations as they are executed.

---

[1] Felix Klein 1908, "Arithmetic, Algebra, Analysis", ISBN 048643480X

Arithmetic

8.1 Equals

Equals is represented by the equal sign =. We are starting from ground zero with the integers, so we want the meaning of = to be straightforward and unambiguous.

Two integer numbers y and z are equal if we verify by counting that y and z are the same number.

For example, y=3+4 and z=1+6 are equal, because we count 3 objects and then 4 more to get 7, whereas we count 1 object and then 6 more and also get 7. This is what 3+4=1+6 means. The equal sign connecting two expressions simply means:

Calculate the values of the integer expressions on both sides of the equal sign by counting to verify that both sides yield the same number.

Take note of the qualifier *integers*. Expressions that include non-integer numbers such as fractions and decimals cannot be evaluated by counting alone.

The basic laws governing the arithmetic operations on integers explain why 17+93 = 93+17, or why (18×3)×6 = 18×(3×6). The point is that two expressions of numbers may be written so that they look different, but are in fact equal.

8.2 Addition

(1) $x + y$ *is always a number* (*addition is always possible*)

(2) $x + y$ *is one valued*

(3) *The associative law holds* $(x + y) + z = x + (y + z)$

(4) *The commutative law holds* $x + y = y + x$

(5) *The monotonic law holds*   *If* $y > z$, *then* $x + y > x + z$

Law (1) means we can always add 1 to get the next number, and that we can add any two numbers to get the sum.

Law 2 means any sum has a unique value.

Expansion of a number into a sum of its parts is clearly valid by the laws (1) and (2). For example from Chapter 2

$$02034 = 00000 + 2000 + 000 + 30 + 4$$
$$56913 = 50000 + 6000 + 900 + 10 + 3$$
the sum is
$$58947 = 50000 + 8000 + 900 + 40 + 7$$

Law (3) states the numbers (x+y)+z and x+(y+z) are equal.

This associative law of addition means *the order of appearance of integers in a finite sum does not affect the sum.*

An expansion of the repeated use of law (3) produces a general statement.

General statement: Given any collection of m numbers $\{n_1, n_2, ..., n_m\}$, addition of the individual numbers in the number $n_1+n_2+...+n_m$ will produce the same result regardless of the order of the addition, and so writing $n_1+n_2+...+n_m$ without parentheses is unambiguous.

Clearly the associative law (3) and commutative law (4) of addition are at work when we manipulate sequences of numbers.

The convention regarding parentheses is this. First execute the operations in the innermost parentheses. Then systematically work your way out by applying the associative and commutative laws.

$$n_1 = ((23+6)+1+14)+9 = (29+1+14)+9 = (30+14)+9 = 44+9 = 53$$

$$n_2 = 23+(6+(1+14+9)) = 23+(6+(14+1+9)) = 23+(6+(14+10))$$
$$= 23+(6+24) = 23+30 = 53$$
$$therefore\ n_1 = n_2$$

Hopefully, it is clear that the associative law (3) and commutative law (4) of addition are labor saving devices. For example there are 120 ways to add five numbers ($5\times4\times3\times2\times1=120$). Now we know *any one* way will do.

Arithmetic

8.3 Multiplication

Reminder of a notation convention: the multiplication sign is omitted when letters are used to represent numbers. And so, xy means x × y.

(6)   *xy is always a number (multiplication is always possible)*
(7)   *xy is one valued*
(8)   *The associative law holds* $(xy)z = x(yz)$
(9)   *The commutative law holds* $xy = yx$
(10) *The monotonic law holds*   *If* $y > z$, *then* $xy > xz$

Law (6) says any product xy is always a number.

Law (7) says any product has a unique value.

Multiply the expanded 9361 by 5 to produce the unique product 5 × 9361.
$$5 \times 9361 = 5 \times 9000 + 5 \times 300 + 5 \times 60 + 5 \times 1$$
$$= 45000 + 1500 + 300 + 5 = 46805$$

An expansion of the repeated use of law (8) produces a general statement.

General statement: Given any collection of m numbers $\{n_1, n_2, ..., n_m\}$, multiplication of the individual numbers in the number $n_1 \times n_2 \times ... \times n_m$ will produce the same result regardless of the order of the multiplication, and so writing $n_1 \times n_2 \times ... \times n_m$ without parentheses is unambiguous.

Clearly the associative law (8) and commutative law (9) of multiplication are at work when we manipulate sequences of numbers.

The convention regarding parentheses is this. First execute the operations in the innermost parentheses. Then systematically work your way out by applying the associative and commutative laws.

$$n_1 = ((23 \times 6) \times 1 \times 14) \times 9 = (138 \times 1 \times 14) \times 9 = (138 \times 14) \times 9 = 1932 \times 9 = 17388$$

$$n_2 = 23 \times (6 \times (1 \times 14 \times 9)) = 23 \times (6 \times (14 \times 9)) = 23 \times (6 \times 126) = 23 \times 756 = 17388$$
$$therefore \ n_1 = n_2$$

The order of appearance of integers in a finite product does not affect the product.

Clearly the associative law (8) and commutative law (9) of multiplication are labor saving devices. For example there are 120 ways to multiply five numbers ($5\times4\times3\times2\times1=120$). Now we know *any one* way will do.

Reminder: use definitions exactly as given.

8.4 Connecting Addition to Multiplication

Multiplication is connected to addition by the distributive law.

(11) $x\times(y+z) = x\times y + x\times z$

or

(11) $x\times y + x\times z = x\times(y+z)$

All laws work in both directions.

Operator precedence: multiply, then add, except for parenthesis.

Law 11 generalizes to any number of numbers. For example

$$n = x(a+b+c+d+e)$$
$$= xa + x(b+c+d+e)$$
$$= xa + xb + x(c+d+e)$$
$$= xa + xb + xc + x(d+e)$$
$$= xa + xb + xc + xd + xe$$

And so $x(a_1 + a_2 + a_3 + \cdots + a_m) = xa_1 + xa_2 + xa_3 + \cdots + xa_m$

Here is an example from Chapter 3

$$375\times9361 = (300+70+5)\times9361 \quad \textit{level 2}$$
$$= (300\times9361) + (70\times9361) + (5\times9361)$$
$$= (3\times9361\times100) + (7\times9361\times10) + (5\times9361\times1) \quad \textit{level 1}$$
$$= 2\,808\,300 + 655\,270 + 46\,805 = 3\,510\,375$$

Arithmetic

8.5 Order

Order answers the question is x less than <, equal to =, or greater than > y? In other words which comes first on a number line.

Here is a geometrical interpretation. Start with positive integers, which increase to the right on the number line.

0 1 2 3 4 5 6 7 8 9 10 11 12 13 14 15 16

Number x is less than y if, when we count up from zero, x appears before y. More formally we say

Integers x and y satisfy x<y when the x's position is to the left of y's position on the number line.

An algebraic interpretation is this.
If x<y, then there is a nonzero integer z so that x+z=y. The two expressions x<y and x+z=y are logically equivalent.

Consequently these statements are true.
If x< y, then there is a nonzero integer z so that x+z=y.
and the converse is
If there is a nonzero integer z so that x+z=y, then x<y.

Substituting specific numbers for x, y, z into these statements, and observe that there is no question the statements are true.

If 5<23, then there is a nonzero integer 18 so that 5+18=23.
and the converse is
If there is a nonzero integer 18 so that 5+18=23, then 5<23.

Observe that z is the number of units from x to y. If x=12 and y=24, starting from 12, takes 12 steps to the right to reach 24.

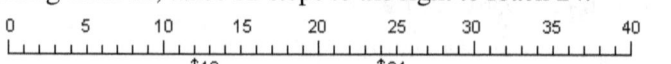
0   5   10   15   20   25   30   35   40
            ↑12   ↑24

Inequalities become even more interesting when negative integers enter the picture. E.g. is −5×2 < −4×3?

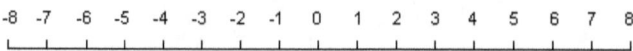
-8 -7 -6 -5 -4 -3 -2 -1 0 1 2 3 4 5 6 7 8

8.6 The Number Line and the $+ - \times \div$ Operations

The number line is a ruler of *infinite* length extending from minus to plus infinity ($-\infty$ to $+\infty$). The line segment from 0 to 1 defines the unit of length. The integers mark a set of equally spaced points defining the infinite ruler. The number line enables a geometric representation of the four operations.

Notation A line segment from point a to point b is represented by [a,b].

Addition

The sum of two integers m and n is represented by joining two line segments of length m and length n. The number line allows for a clear presentation of the commutative law m+n=n+m.

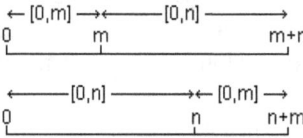

Subtraction

If m, n are integers and m < n, then the difference of two integers m and n is represented by including a line segment of length m "inside" a line segment of length n. The associated law (n−m)+m =m+(n−m) is also shown clearly on the number line.

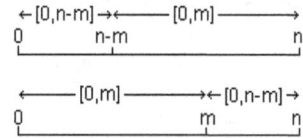

Multiplication

The product of two integers m and n is shown on the number line by adding a new scale to the number line. Here is 3×7=21.

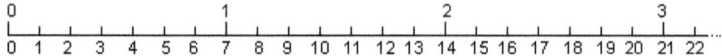

The commutative law mn=nm is clearly demonstrated on the number line. Here is 7×3=21. Compare to 3×7=21

Arithmetic

Areas Areas are another way to show products. A square has sides of length 1, so that its area is 1×1 = 1. And, here are area representations of 3×4=12, and 4×3=12. Clearly 4×3=3×4. The expression 4×3=3×4 also can be "proven true" by counting the number of unit squares on the left side, and by counting the number of unit squares on the right side.

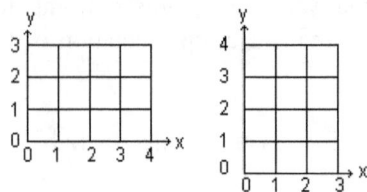

Remark Saying that the area of the unit square is the number 1 is simply a convention. If the unit square is different from 1, e.g. 3, then the resulting expressions become complex. So mathematicians use 1.

Division

The division of two integers m and n is shown on the number line by marking off multiples of the quotient q. Here is 33÷7=4+(5÷7), or 33=7×4+5, and x÷y=q+(r÷y) , or x=y×q+r, or x=qy+r.

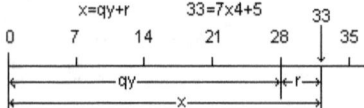
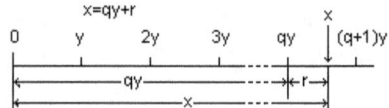

116

Answers to the Chapter 6 and 7 Problems

Problems 6 Fractions

An object is divided into 23 parts.
1. What is one part called? one twenty third
2. How is it written?   1/23

A number = 23/145.
3. Into how many parts is the original object divided? 145
4. How many parts of the original object does the fraction represent? 23

Find fractions equivalent to the following fractions that have a denominator of 128

5. 1/4	6. 3/8	7. 7/16	8. 5/32	9. 17/64
32/128	48/128	56/128	20/128	34/128

Find fractions equivalent to the following pairs of fractions that have a minimum common denominator

10. $\dfrac{1}{4}$ $\dfrac{3}{5}$	11. $\dfrac{1}{14}$ $\dfrac{3}{21}$	12. $\dfrac{7}{15}$ $\dfrac{5}{12}$	13. $\dfrac{11}{16}$ $\dfrac{3}{15}$	14. $\dfrac{1}{7}$ $\dfrac{3}{8}$
5/20 12/20	3/42 6/42	28/60 25/60	165/240 48/240	8/56 21/56

For each fraction find quotient plus remainder fraction.

15.	16.	17.	18.	19.	20.	21.	22.
8/8	9/8	10/8	11/8	12/8	13/8	14/8	15/8
1+0/8	1+1/8	1+2/8	1+3/8	1+4/8	1+5/8	1+6/8	1+7/8

23. 42/7	24. 22/7	25. 12/3	26. 14/3	27. 30/6	28. 33/6	29. 65/21
6+0/7	3+1/7	4+0/3	4+2/3	5+0/6	6+3/6	3+2/21

30. 27/11	31. 15/11	32. 19/4	33. 23/14	34. 45/9	35. 53/9	36. 27/5
2+5/11	1+4/11	4+3/4	1+9/14	5+0/9	5+8/9	5+2/5

Change to higher terms.

37. 2/7 to 28ths	38. 3/5 to 20ths	39. 9/8 to 32ths	40. 3/10 to 1000ths
8/28	12/20	36/32	300/1000

41. 11/6 to 72ths	42. 7/13 to 65ths	43. 2/9 to 27ths	44. 23/12 to 24ths
132/72	35/65	6/27	46/24

Arithmetic

Find the missing numbers.

45.	46.	47.	48.	49.	50.	51.
$\dfrac{42}{7}=\dfrac{}{28}$	$\dfrac{2}{3}=\dfrac{}{27}$	$\dfrac{9}{4}=\dfrac{36}{}$	$\dfrac{}{6}=\dfrac{25}{30}$	$\dfrac{30}{}=\dfrac{5}{3}$	$\dfrac{33}{7}=\dfrac{3}{}$	$\dfrac{8}{21}=\dfrac{48}{}$
168	18	16	5	18	77	126

Add the following pairs of fractions.

52.	53.	54.	55.	56.	57.
$\dfrac{8}{9}+\dfrac{1}{5}$	$\dfrac{3}{7}+\dfrac{1}{4}$	$\dfrac{5}{9}+\dfrac{3}{5}$	$\dfrac{9}{13}+\dfrac{2}{39}$	$\dfrac{2}{21}+\dfrac{3}{7}$	$\dfrac{4}{14}+\dfrac{1}{3}$
49/45	19/28	52/45	29/39	11/21	26/42

Convert to fractions and add.

58.	59.	60.	61.	62.	63.
$2\dfrac{7}{21}+4\dfrac{5}{14}$	$3\dfrac{1}{12}+5\dfrac{3}{18}$	$1\dfrac{3}{8}+3\dfrac{5}{6}$	$6\dfrac{7}{13}+1\dfrac{3}{26}$	$\dfrac{7}{16}+1\dfrac{11}{12}$	$4\dfrac{3}{11}+1\dfrac{2}{9}$
281/42	297/36	125/24	199/26	113/48	544/99

Find the gcd of each pair of numbers.

64.	65.	66.	67.
255 *and* 153	336 *and* 280	136 *and* 255	105 *and* 168
51	56	17	21

Find the lcm of each pair of numbers.

68.	69.	70.	71.
255 *and* 153	336 *and* 280	136 *and* 255	105 *and* 168
765	1680	2040	840

Add the following pairs of fractions using the lcm

72.	73	74.	75
$\dfrac{13}{255}+\dfrac{7}{17}$	$\dfrac{4}{255}+\dfrac{5}{153}$	$\dfrac{9}{336}+\dfrac{2}{7}$	$\dfrac{13}{336}+\dfrac{3}{140}$
118/255	(12+25)/15×17×3	(9+96)/336	(65+36)/28×60

Find the lcm of

76.	77.	78.	79.
$\dfrac{1}{4}\ \dfrac{3}{5}\ \dfrac{7}{9}$	$\dfrac{2}{3}\ \dfrac{1}{15}\ \dfrac{9}{10}$	$\dfrac{5}{6}\ \dfrac{1}{2}\ \dfrac{1}{36}$	$\dfrac{3}{8}\ \dfrac{2}{9}\ \dfrac{1}{5}$
180	30	36	360

Subtract the fractions using lcm

80.	81.	82.	83.
$\dfrac{13}{85} - \dfrac{1}{17}$	$\dfrac{37}{51} - \dfrac{5}{255}$	$\dfrac{1}{14} - \dfrac{7}{168}$	$\dfrac{13}{56} - \dfrac{3}{140}$
8/85	180/255	5/168	59/280

84. Decrease the value of 3/4 by twelve thirty ninths. 69/156
85. Decrease the value of 7/10 by two fifths.   3/10
86. Decrease the value of 4/9 by one fifth.   11/45
87. Decrease the value of 3/7 by one tenth.   23/70
88. Decrease the value of 17/32 by three eighths. 5/32
89. Decrease the value of 5/16 by two sevenths.   3/112

Multiply the fractions. Reduce to lowest terms.

90.	91.	92.	93
$\dfrac{3}{22} \cdot \dfrac{55}{17}$	$\dfrac{18}{34} \cdot \dfrac{51}{28}$	$\dfrac{84}{13} \cdot \dfrac{39}{132}$	$\dfrac{28}{57} \cdot \dfrac{76}{7}$
15/34	27/28	21/11	304/57

94. Increase the value of 3/4 by four times.   3/1
95. Increase the value of 3/10 by two and one-half times.3/4
96. Increase the value of 4/9 by five times.   20/9
97. Increase the value of 17/32 by two and two thirds times. 17/12
98. Increase the value of 4/9 by four and one half times.   2/1
99. Increase the value of 12/19 by one and one half times. 18/19
100. Decrease 1/3 to 1/7.   1/3 times 3/7
101. Decrease 3/5 to 1/10.   3/5 times 1/6
102. Decrease 17/32 to 1/16.   17/32 times 2/17
103. Decrease 53/120 to 1/3.   53/120 times 40/53
104. Decrease 5/9 to 5/13.   5/9 times 9/13
105. Decrease 2/7 to 3/100.   2/7 times 21/200
106. Find 1/3 of 1/9   1/27
107. Find 2/7 of 9/11   18/77

Divide the fractions. Reduce to lowest terms.

108.	109.	110.	111.
$\dfrac{\frac{3}{22}}{\frac{17}{55}}$	$\dfrac{\frac{18}{34}}{\frac{28}{51}}$	$\dfrac{\frac{13}{84}}{\frac{39}{132}}$	$\dfrac{\frac{57}{28}}{\frac{76}{7}}$
15/34	27/28	11/21	57/304

Arithmetic

Problems 7 Decimals

1. The denominator of a Decimal fraction is a power of what number? 10
2. What is the purpose of the Decimal point?
   to mark separation of integer and fraction parts
3. A mixed number is the sum of two parts. What are their names?
   integer, fraction
4. The decimal point separates two parts of any number. What are their names?
   integer, decimal fraction

Convert to a decimal.

5.	6.	7.	8.	9.	10.
$\dfrac{3}{6}$	$\dfrac{3}{7}$	$\dfrac{5}{20}$	$\dfrac{13}{18}$	$\dfrac{17}{123}$	$\dfrac{5}{16}$
0.5	0.428571	0.25	0.7222	0.13821	0.3125

Convert to a decimal.

11.	12.	13.	14.	15.
$\dfrac{1}{6}$	$\dfrac{1}{11}$	$\dfrac{13}{45}$	$\dfrac{23}{54}$	$\dfrac{45}{111}$
0.1666	0.090909	0.2888	0.4259259	0.405405...

Convert the following phrases into decimals.
16. One-hundred-three one-thousandths 0.103
1. Thirty-nine one-thousandths   0.039
18. Nine one-hundredths   0.09

Expand the numbers into a sum of powers of ten. Omit the zero terms.

19.	20.	21.	22.
5.3	347.59	502.01	20001.0001

19. $5 \times 10^0 + 3 \times 10^{-1}$
20. $3 \times 10^2 + 4 \times 10^1 + 7 \times 10^0 + 5 \times 10^{-1} + 9 \times 10^{-2}$
21. $5 \times 10^2 + 2 \times 10^0 + 1 \times 10^{-2}$
22. $2 \times 10^4 + 1 \times 10^0 + 1 \times 10^{-4}$

23. What operation with what number moves a decimal point to the left?
   divide by ten

24. What operation with what number moves a decimal point to the right?
multiply by ten

25. Adding a leading zero to the integer part with value x changes the value to what?   same number x

26. Adding a trailing zero to the integer part with value x changes the value to what?   10x

27. Adding a leading zero to the fractional part with value x changes the value to what?   x/10

28. Adding a trailing zero to the fractional part with value x changes the value to what?   x no change

29. Write thirty seven hundredths in numerical format.   0.37
30. Write thirteen tenths in numerical format.   1.3
31. Write fifty seven thousandths in numerical format.   0.057
32. Write one hundred ninety nine ten thousandths in numerical format
        0.0199
33. Write one one millionth in numerical format.   0.000001
34. Find the repeating decimal produced by 1/6.   0.166666...
35. Find the repeating decimal produced by 1/11.   0.090909...
36. Find the repeating decimal produced by 2/13.   0.153846153846...
37. Find the repeating decimal produced by 77/123.   0.6260162601...
38. Find the repeating decimal produced by 2/11.   0.181818...
39. Find the repeating decimal produced by 12/123.   0.09756097560...

Use a calculator.
40. Find the non-repeating decimal produced by square root of 3.
        1.73205080...
41. Find the non-repeating decimal produced by p/2.   1.570796327...
42. Find the non-repeating decimal produced by log 2   0.301029996...
43. Find the non-repeating decimal produced by ln 2   0.693147181...
44. Find the non-repeating decimal produced by e^2 7.389056099...
45. Explain why the ratio of two integers cannot be irrational.

By definition they are rational. As a practical matter you can mark the point on the number line. E.g. 12/123. Divide distance from 0 to 1 into 123 parts, count 12 parts to right from 0.

Arithmetic

Convert to percent to three decimal places.

46.	47.	48.	49.	50.	51.
$\dfrac{1}{4}$	$\dfrac{3}{25}$	$\dfrac{33}{100}$	$\dfrac{12}{20}$	$\dfrac{1}{14}$	$\dfrac{3}{9}$
25.000%	12.000%	33.000%	60.000%	7.143%	33.333%

Find the specified percent of numbers to three decimal places.

52.	53.	54.
22%of 123	7%of 9980	39%of 243
27.060	698.600	94.770

Find the sums and differences.

55.	56.	57.	58.
23.005+322.1	23.005−322.1	3007.1+1.3002	3007.1−1.3002
345.105	−299.095	3008.4002	3005.7998

59.	60.	61.	62.
100.01+1.01	100.01−1.01	5.06+3.067	5.06−3.067
101.02	99.00	8.127	1.993

63.	64.
90034.002+100.6	90034.002−100.6
90134.602	89933.402

Find the products.

65.	66.	67.	68.	69.
23.005×322.1	3007.1×1.3002	100.01×1.01	5.06×3.067	90034.002×100.6
7409.91050	3909.831420	101.01010	15.519020	9057420.601

70.	71.	72.	73.	74.
23.005÷322.1	3007.1÷1.3002	100.01÷1.01	5.06÷3.067	90034.002÷100.6

70. q = 0, r = 0.0714211919
71. q = 2312, r = 0.798031
72. q = 99, r = 0.019801980
73. q = 1, r = 0.649820672
74. q = 894, r = 0.9701988

Index